THE TRUE COST OF WOOL

A Vision for Revitalizing the Canadian Industry

ANNA HUNTER

9/10 PUBLICATIONS

VANCOUVER

3rd Printing, February 2026

Nine Ten Publications Inc.
Vancouver
shop.ninetenpublications.ca

Library and Archives Canada Cataloguing in Publication

Title: The true cost of wool : a vision for revitalizing the Canadian industry / Anna Hunter.

Names: Hunter, Anna, author.
Description: Includes bibliographical references.
Identifiers: Canadiana (print) 20250196360 | Canadiana (ebook) 2025019645X | ISBN 9781738360710 (softcover) | ISBN 9781738360727 (EPUB)
Subjects: LCSH: Wool industry—Canada. | LCSH: Woolen goods industry—Canada. | LCSH: Wool—Canada.
Classification: LCC HD9904.C32 H867 2025 | DDC 338.4/7677310971—dc23

Editing by Kim Werker
Copyediting and fact checking by Michelle Woodvine

Design by Ninth and May Design Co.

Production of this book was made possible by the enthusiasm and financial support of readers and wool lovers who pre-ordered the book long before it came out. Thank you for making it possible for this book to be made!

Visit shop.ninetenpublications.ca for more books and booklets about how to make things, how things are made, and where our art and craft materials come from.

Printed and bound in Canada on 100% post-consumer waste recycled Sustana Enviro paper, which is processed chlorine-free. It is designated Ancient Forest Friendly™ and FSC® certified.

PERMANENT

PCF

This book is dedicated to those
who imagine a better world is possible,
and then roll up their sleeves to make it happen.

LAND ACKNOWLEDGEMENT

I am currently a settler on Treaty 1 territory—the land that I farm, steward, live on and depend on has been home to Indigenous Nations since time immemorial. Across this land Indigenous People have long nurtured deep, reciprocal relationships with the soil, water, plants, and animals that have sustained ecosystems and communities.

I can't share the history of wool in Canada without acknowledging that sheep and wool production are deeply woven into the colonization of this land and the theft of Indigenous land and sovereignty.

Sheep were brought to these lands as part of the settler colonial agricultural systems that dispossessed Indigenous People of their territories. The Canadian Department of Militia and Defence worked with Canadian mills to produce and procure wool tunics for the soldiers that took part in the violent suppression of the Métis resistance in 1885.

The Canadian wool industry has been, in part, built on the agricultural theft of Indigenous lands. It is our responsibility to hold this truth alongside our aspirations for a new sheep and wool industry. As we continue to work with wool, we must prioritize reconciliation that is rooted in accountability, relationship building, and action. My hope is that we move forward in a spirit of collaboration, gratitude, and respect for this land and an ongoing commitment to true reconciliation.

TABLE OF CONTENTS

FOREWORD

From apples to flip flops to that beautiful skein of yarn you bought last week, it's increasingly challenging to figure out where things came from. All too often, an opaque global supply chain ends up blurring the true material, environmental, and human costs of what you're buying. And without greater visibility into the supply chain, how are we supposed to make informed buying decisions, much less help others do the same?

In our yarn world, the biggest factor is fiber content. At one end of the spectrum are fibers made from virgin petroleum and chemically regenerated cellulose (i.e., polyester or viscose bamboo). They're manufactured by a handful of global petrochemical companies, and their low price is often subsidized by economies of scale, lax environmental regulations, the exploitation of a cheap workforce, and generous tax and regulatory loopholes. Synthetic fibers take from a finite supply of virgin petroleum and return only fodder for the landfill.

At the other end of the spectrum we have the miracle that is wool, which doesn't come from a well, barrel, tub, or spigot. Its production isn't turned on and off whenever market forces dictate. The industry isn't dominated by a handful of monopolies. In fact, all you really need is some land and water and sunshine. With a little planning, you could go from lamb to sweater without ever leaving home.

Where synthetic fibers are extractive, wool is inherently regenerative. The fiber grows from the follicles of a sheep, 365 days a year, whether we want it to or not. What wool takes in grass or grain or water or energy it more than replenishes with increased soil fertility, carbon sequestration, and a healthy, long-wearing material that requires little laundering, releases zero microplastics, and is fully biodegradable at the end of its functional lifespan. And that's not even mentioning the milk, meat, and skin that sheep offer up at the end of their *own* lifespan.

We're lucky to be able to raise sheep pretty much anywhere in the world—but they won't be the same breed, and they won't all grow the same kind of wool. In a global market that only prizes certain kinds of wool, this puts many countries at a disadvantage.

What happens in places like Canada, whose dominant wool production doesn't hold high value on the commercial market? Should those producers simply close up shop and go gently into that good night? Of course not. But how do you turn things around?

For first-generation Canadian shepherdess and mill owner Anna Hunter, it keeps coming back to one statistic in particular: More than 90% of Canada's wool is exported. Meanwhile, more than 95% of the wool yarn that Canadians consume is imported from other countries.

This makes no sense—not from a commercial perspective, not from a logistical perspective, and certainly not from an environmental perspective.

How did we get here, where *are* we, exactly, and how can we turn Canadian wool around?

Anna has been on a journey for answers. She is a fierce champion for Canadian wool. Her knowledge is deep, her passion infectious.

This book is wool geekery at its finest, and I'm all here for it. The fact that you're reading these words tells me you, too, are all here for it—that you're ready to go deeper, learn more, and spread the word.

Let's do this!

Clara Parkes
Brooksville, Maine
April 2024

AUTHOR'S NOTE

I submitted the first draft of this book ten days after Donald Trump was elected president of the United States on November 5, 2024. I had no idea how timely my book would become as his second term in office would begin to unfold just a couple of months later.

As his plans for a trade war with Canada became known, yarn crafters took to every social media platform hungrily searching for "Canadian wool," and tagging indie yarn dyers and local yarn shops to promote our Canadian wool industry. Many of us who have been working to build up the local wool industry began to explain the nuances between yarn "dyed in Canada," "distributed by a Canadian company," or "sold in a Canadian yarn shop" but not actually manufactured in Canada or made of Canadian wool. This book about the Canadian wool supply chain is suddenly even more relevant and timely given the strange situation we find ourselves in at the time of publication in the spring of 2025.

It is not yet clear how tariffs will affect the wool industry in Canada, but here's what we do know. We know that about $1.3 million of raw wool was sold to the U.S. in 2022.[1] And, while we don't have hard data about how much finished or value-added wool products are exported from Canada to the U.S., we know this trade war will impact our industry one way or another.

My hope is that we will take our cue from our nineteenth-century Canadian government and wool manufacturers, who saw the failure of the Canadian-American Reciprocity Treaty (see page 62) as an opportunity to innovate and invest in our domestic industry, with Canadian mill owners using wool no longer being exported to American mills to produce tweed and flannel fabric, leading to a boom in manufacturing and pricing for farmers.

---

1 Ashley Bacon, "Ontario wool producers focused on the positive amid U.S. tariff threat," *CTV News*, March 6, 2025, https://www.ctvnews.ca/kitchener/article/ontario-wool-producers-focused-on-the-positive-amid-us-tariff-threat/

Now is the time to build a system of production that prioritizes high-quality products, integrated supply management between farmers and manufacturers, and de-carbonized methods of wool transportation and production. A system that provides meaningful employment and living wages, and invests in education and training within the textile manufacturing industry. We need a system of production that can adapt to our changing climate and political circumstances while maintaining community-based, regional approaches to wool textile supply chains.

This new system needs to be built for the long-term survival of planet and people, not as a retaliatory measure. My call for building up domestic infrastructure is not a call to reinforce borders or celebrate a nationalism that hasn't yet reckoned with its devastating colonial history. It's a call based on the belief that regionalism is an antidote to climate change and is an ecological necessity, that we must value the rights and standards of labourers regardless of where they live and work, and that our society must strive to provide meaningful employment in rural Canadian communities. This is a system that does not prioritize profit over the impact on land, animal, and people.

As an advocate working tirelessly to build up our local Canadian wool industry and call out the problems of fast fashion, I do not align myself with the profit-centric and protectionist tariff battle cry coming from the south. I see the solutions I lay out in this book as long-term investments in our communities. I see them as opportunities to say our ideas out loud (even if they seem impossible) and as invitations to work together to find ways to make them real. Globalized supply chains and their inevitable consequences—environmental destruction and labour exploitation—don't have to be our only option. Another world is possible, and I hope that you will join me and take up the call to make it happen, to bring our efforts together during this time of uncertainty and chart a different path forward.

April 8, 2025

Introduction

HOW DO WE CHOOSE YARN OR WOOL PRODUCTS?

If you are a yarn crafter like me, chances are when you travel to any new city, town, or region one of the first things you look up is where the yarn stores are. Maybe you are obvious about it and you plan a full yarn crawl with your crafty friends and map out the order in which you will hit up the shops to ensure maximum enjoyment and shopping.

Or perhaps you are a bit sneaky like me, and figure out how close each yarn shop is to a great restaurant or museum so your kids and spouse don't know that you actually just convinced them to drive an extra thirty minutes to eat at the cool lunch spot and then, "Oh wow, look at that, there is a yarn store across the street—I'm just going to pop in for a few minutes!"

We've all been there: the eager anticipation of a new yarn shop, the wall-to-wall cubbies of yarn and needles and hooks and possibilities.

What do you notice first when you walk into those shops? Do you beeline to a brand you haven't seen before in person? Do you look for a specific colour—solid, semi-solid, tonal, variegated, self-striping, natural shade? Do you shop with a specific project in mind, demonstrating self control to limit your purchase to what you already have a plan for?

Do you pick up the yarn, pull it to your nose and take a big, long breath in and smell? Do you rub it against your cheek or your neck and see how it feels next to your skin? Do you hold it next to your face in a mirror to see if the colour will work with your hair or complexion?

Do you walk instinctively toward the window for natural light to make sure the colour looks as good as it does in the store's lighting?

Do you turn the ball or hank or skein over in your hand and look at the meterage? Do you look for the weight or gauge or the quirky name given to this particular speckled or neon or fractal dyed yarn? Maybe this is where you look at the price, and do the mental gymnastics to see if a sweater's worth fits into your yarn budget this month?

Maybe you look for where the yarn came from. Was it dyed in Canada? Is it from the Shetland Islands, or Iceland, or maybe the Peruvian Highlands? Do you ask yourself what the farm was like where the natural fibres comprising this yarn originated? Do you picture idyllic rolling green hills with happy sheep grazing on grass and lazily looking toward the horizon? Do you picture the rough hands—with dirt and lanolin under the fingernails—of the shearer who removed the wool from the sheep? Can you smell the hot and humid air in the mill where the wool was scoured, carded, spun, plied, and skeined? Do you picture the large vats of dye and water that the yarn sits in while it soaks up the synthetic or natural shades applied to it? Can you see it bagged and tagged and shipped and unpacked and put on shelves? Can you envision the full life cycle of the yarn in your hand?

Maybe you are not a crafter, but someone who wants to make informed decisions about the clothing you buy. Most of the questions I asked above apply to clothing as well as yarn. Perhaps when you travel you seek out locally made clothes or home goods. Or when you are at home you want to support local clothing manufacturers who use textiles produced in your region.

This book focuses specifically on wool, which is the textile fibre grown by sheep. Most yarn and clothing labels offer little information about where or how the wool yarn or product was produced or the people involved in its creation. Our wool supply chains are complex, which makes it challenging to distinguish the source of our wool and the process it goes through to get to us. Without accurate, easily accessible, transparent information about how and where our wool is processed we cannot make informed decisions about what we buy. This book is for those who want to dig a little deeper and understand the wool supply chain in Canada and beyond.

WHAT DOES YARN COST?

I operate a sheep farm and wool mill. Many of the customers who purchase the yarn I produce started crafting during the COVID-19 pandemic. During that tumultuous time many people turned to their needles, hooks, loom, or spinning wheel as a source of comfort, distraction and peace. But the pandemic also had an impact on the wool supply chain. The global wool market shut down as the world's largest wool manufacturing nation (China) halted production. The sale of Canadian raw wool, most of which is shipped out of the country, ground to a halt and wool began to stockpile in warehouses as uncertainty over the future of the industry crept in.[1]

This disruption gave us a great opportunity to critically examine the fragile supply chain and our long-term access to wool not only as a material for our favourite hobby but as a critical material in our textile and clothing systems. My hope at the time was that a new conversation would emerge about the source of our wool, and that we would collectively commit to reimagining a value chain and a new system for producing our textiles and materials: A system that would prioritize community economic development, environmental resiliency, and climate-positive growth. A system that would prioritize ecological and social health above the bottom line and infinite profit margins.

Our current system of wool-material and textile manufacturing—both for the clothing we wear and yarn we craft with—is a globalized model that involves the production of wool goods being outsourced to low-cost economies, particularly those where wages are low and environmental oversight lacks rigour. This outsourcing drives down production costs, making wool products cheaper to produce and increasing profits for those at the top of the wool value chain. However,

1 Eric Bjersgo, "Market Report," *Canadian Cooperative Wool Growers*, Fall 2020, https://wool.ca/page/wool-market-reports.

as production becomes more removed from local communities, we lose transparency into how our goods are made, limiting our ability to make informed and conscious consumer choices.[2]

Would we make different choices if there was more transparent information on our yarn and clothing labels? Would we pause if the label stated that the material had travelled more than 13,000 km throughout its processing life? Would we reconsider if we knew the mill workers only made the equivalent of $2.15 per hour?[3] Or if the label indicated what waterway the superwash effluent or dye ended up in?[4] Would things be different if the true costs were boldly printed on our yarn and clothing in the same way the colour, brand, and meterage is displayed?

So what does it really cost to produce our yarn? What is the cost of raising a sheep for a year? How much do mills charge for processing wool? How much does it actually cost to run a wool-based business in Canada? I want to take the conversation even further than that, to uncover the larger costs—the hidden costs—in the yarn and wool that we use.

"Why Is Your Yarn So Expensive?"

At the start of 2019, when my wool mill and sheep farm were still in their early stages, I spent a lot of time in the mill. Like, sixteen-hour days in the mill, just trying to figure out how to make a quality product with a dynamic resource like wool, and with very little formal training, the learning curve was steep. That year was a hard one. One night, at

2 Isabel Slone, "Why isn't more clothing made in Canada?" *Maclean's*, July 21, 2021, https://macleans.ca/society/why-isnt-more-clothing-made-in-canada/.

3 I calculated this number by looking at the minimum wage in Shanghai (although minimum wage is set regionally in China). I took the full minimum wage per month of $345 CAD and divided it by 140, which is 8 hours a day for 5 days a week for 4 weeks a month. (ILO, January 2023.)

4 The most common form of superwashing wool yarn is the Hercosett-Chlorine method. Each wool fibre is covered in thousands of overlapping scales—these scales are what cause felting. Superwashing wool combines a process of stripping off the scales using a chlorine solution then coating the fibres in polyamide-epichlorohydrin resin (Hercosett 125) to flatten the remaining scales. (R.R.D. Holt, "Introduction to Superwash Wool," *Journal of the Society of Dyers and Colourists*, 91: 38-44, 1975, https://doi.org/10.1111/j.1478-4408.1975.tb03228.x).

about quarter to nine, I tucked the kids into bed and came back out to the mill to finish spinning some Tunis yarn from one of my favourite local farms. At the same time, I was rinsing yarn I had just dyed and skeining some beautiful alpaca that was ready to go back to a local farmer. I opened up my Instagram to post a photo of the yarn and there was a message waiting for me. Someone I hardly knew had responded to a sales post about our new yarn with a comment about how expensive my wool was and they wondered why my prices weren't comparable to those of the retail store Michaels.

This wasn't the first time I had heard that critique, but somehow it still bothered me. I wrote a response—and deleted it—and wrote it again—rubbed my weary shoulders and realized it was probably best if I walked away. I put my phone down and started rage-cleaning the mill while writing my response in my head. I wanted to ask her if she had any idea what went into making this yarn? The work of being a shepherd, the challenges of starting and running a small-scale manufacturing mill, trying to make it alongside the giants in the industry. My mental first draft ended with a dramatic ALL CAPS comment that was something like "The CHEAP YARN you are referencing doesn't require any less resources or time—it just means you are not paying the TRUE COST of that yarn, and someone else, or something else, is subsidizing it for you."

I wrote "True Cost of Wool" on a little sticky note and stuck it to the cone winder in the mill. That note, and the feeling I had that night, remind me that so many of us don't realize what it actually takes to make yarn or grow wool. We are disconnected from the source of our materials and the process of transforming it. This book, my teaching, and other advocacy work is focused on shining a light on the true cost of our wool and connecting each of us to the many hands behind our yarn. My hope is that by building stronger connections we will be inspired to work collectively to change the industry, to value and account for those actual costs, and to build a just and equitable world that is, of course, covered in wool!

How Do We Determine the True Cost?

As I started researching the true cost of wool, I looked at our food and agriculture systems, and kept coming across the term "True Cost Accounting," an accounting approach that accommodates the impact of our economic activities on environment, society, and health. The Sustainable Food Trust suggests that we pay for our food in a variety of ways beyond just the price at the checkout.[5] We can and should assess the true costs and benefits of our production systems and measure the harmful impacts on the environment, animal welfare and public health. Throughout this book I attempt to paint a full picture of the various full costs of our system of wool production—both domestic and international. I have tried my best to be thorough in my research and to present the "true costs," including labour costs, and costs to both the environment and community. I have occasionally relied on existing bodies of research for our agri-food systems as there isn't dedicated research within the wool and fibre community in Canada.

FOLLOW THE WOOL

"The first rule of Sheep Club is that we don't talk about Sheep Club." This became a running joke in our house, the livestock version of the famous line from the 1999 movie *Fight Club*.

Sheep Club is actually a monthly gathering of sheep farmers hosted by the Manitoba Sheep Association in every region throughout Manitoba. I have attended many of the Eastern Manitoba Sheep Clubs and in the early days of my farming journey I appreciated the camaraderie and the advice. One of the things I learned very early on in Sheep Club is that my sheep farm and my focus on wool puts me in the minority here in Canada—most shepherds raise sheep for meat. In fact, wool is often described as a by-product of the lamb industry and in the commercial farming industry wool is generally an after-thought. I was overwhelmed and out of my depth with the early talk of "feed-conversion-rate" and

5 Sustainable Food Trust, https://sustainablefoodtrust.org/our-work/true-cost-accounting/.

the variable rate for market lambs and genetic testing for scrapie resistance. If you have no idea what I'm talking about then you are in good company. I just wanted to talk about wool quality!

In those early days, I had to work through my imposter syndrome in sheep farming spaces—and I also had a lot to learn from the farmers in my community about how interconnected wool quality and things like "feed-conversion-rate"[6] are. But I also learned that many Canadian sheep farmers have become disillusioned with the idea that wool might someday be worth something, and if I wanted to impact the industry for good, I'd need to have "boots on the ground" (or "boots in the mud" if we're talking springtime in Manitoba). Ultimately, we need more wool people in sheep farming spaces, building connections and creatively working on wool together.

My feelings of inadequacy at my first Sheep Club meetings were in stark contrast to the first fibre art show that I attended after starting my farm. I had only had sheep for six weeks at that point, but everyone who came to my booth looked at me as though I had done the impossible—gone from knitter to sheep farmer—and I was now the expert on the latter. No one asked me about feed-conversion-rates, and I wore the sheep manure that was still on my jeans like a badge of honour.

In that first year, I realized that I was in a unique position to bridge the gap between the world of sheep farming and the world of knitters and fibre artists. I could attend sheep farming conferences and I could also talk about Ravelry and knitting patterns. Perhaps this in-between position in both of these unique communities could support exactly the kind of creative and innovative approach the wool industry needs.

In 2023 I invited a sheep specialist from Manitoba Agriculture to attend the Manitoba Fibre Festival. I could tell that he was a bit hesitant about showing up. I'm pretty sure he imagined the event would happen in a small community centre with a couple of ladies sitting around knitting. But I was persistent, and he showed up...and stayed for almost four hours. He was flabbergasted at the general excitement in the build-

6 Feed conversion rate measures the efficiency that livestock (in this case sheep) convert feed into meat or milk or wool.

ing, the number of shoppers, the amount of money being spent on wool (of all things) and the interest in the agriculture demonstrations. He remarked that we "really have something incredible here that the sheep industry needs to witness." He couldn't have been more right.

Follow the Wool

To paint the full picture of wool in Canada it is important to first distinguish between the different types of sheep farms and the different types of wool they are producing. My farm, and others like it, are raising flocks of sheep that consist of specific wool breeds—throughout the book I will refer to them as *fibre farmers*. This means wool production and quality is the priority in our livestock management, our main revenue source is the wool, and we generally market our wool products directly to customers through a value chain process—think yarn, roving, wool bedding, wool pellet fertilizer, et cetera.

Unlike fibre farmers, most Canadian sheep farmers are focused on the production of lamb meat, and they sell it through the domestic and international markets—throughout the book I will refer to them as *commercial farmers*. The goal of commercial sheep operations is to produce the highest quality lamb or breeding stock. Some of these operations also prioritize wool production, while others have chosen breeds that don't produce wool at all (these are considered *hair breeds* and they shed their coats every year. Because it is hair and not wool it cannot be used in textiles and yarn). Some commercial operations focus on breeds that are considered dual purpose—they grow quality wool and are also fast-growing for the lamb industry or they produce milk.

In this book, I take you on a journey through these two distinct value-chain paths. We will **"follow the wool"** from sheep to finished goods, identifying and building a framework for the true costs of the wool grown and consumed in Canada.

Throughout the next few chapters you will find the consolidated data in "Follow the Wool" boxes, infographics, and tables. The costs listed are as accurate and precise as possible, considering that markets change, that geographical regions determine disparities in market price for hay and other inputs, and that each operation is unique and thus their costs and revenue will vary. Although these numbers change over time, I believe that they provide us with an important starting point—a snapshot of where the wool industry is now—as well as a roadmap for where we want to go. My hope is that one day we can look back at these numbers and see how far we have come.

My goal in writing the *True Cost of Wool* is to provide you, as both reader and consumer, with more information about how and where your wool comes from so you can make informed choices about what you consume. I hope the examples this book will inspire you to take action to support our local wool community and contribute to the momentum of systemic change in the way we value and manage our domestic wool supply.

> *My hope is that by building stronger connections we will be inspired to work collectively to change the industry, to value and account for those actual costs, and to build a just and equitable world that is, of course, covered in wool!*

Glossary of Terms

There are some words and concepts that I use interchangeably throughout this book. Here are the definitions I work with.

Sheep producer or sheep farmer. I use these two terms to mean the same thing: someone involved in the raising and breeding of domestic sheep. The term "producer" seems to be used more in the large-scale context, but both are appropriate labels.

Value chain and supply chain. The term "value chain" describes the process of adding value to raw materials through production and manufacturing to create a finished product. "Supply chain" refers to the steps taken for a product to get to an end-user.

Regenerative farming. This is not a static definition but an overall belief that agricultural activities should work toward regenerating the health of soil, increase biodiversity, and work in cooperation with the natural water and mineral cycles of a given bioregion. Regenerative farming promotes strategies to reach these goals, including crop diversity, minimal soil disturbance, integrated livestock and crops, maintaining living roots year-round, keeping soil covered, and more.

Traceability. The ability to track every aspect of a product's life cycle. For wool, this includes information regarding the farm it was grown on, the manufacturing of the wool into a material like yarn, batting, or felt, and the distribution of the end product.

Wool Clip. The total amount of wool sheared from a particular farm, or in a specific region, over a set amount of time (e.g., the Manitoba wool clip in 2024). It is used as an aggregate term.

CHAPTER 1

Growing Wool

STATE OF THE INDUSTRY

"There are sheep in Canada?! I thought Merino was only grown in Australia."

I often get this comment when I mention to a knitter that I raise sheep. I find it interesting for two reasons. First, Merino has become synonymous with wool in general for many knitters. (There are over 200 domesticated breeds of sheep, Merino being just one of them!). Second, Canada does indeed have a thriving sheep industry—one which produces a diverse number of sheep breeds, many of which can be used to make excellent fibre products of the same quality that a knitter who has been conditioned to seek out Merino would enjoy.

Whether or not I've had my coffee yet often determines the intensity of my response to this question.

If you catch me early, I might just smile and nod and say, "It's true! We do have sheep in Canada!" Then give a quick response about the variety of breeds that can be found from coast to coast to coast.

But if I'm really wired, look out. I'm about to download the entire history of wool production in Canada on you, taking it all the way up to the current state of the industry. All in one go.

Which, in a nutshell, is what I hope to accomplish with this book. However, the difference between putting it in print versus me ranting all this to a knitter at a stitch 'n' bitch means I can pause for paragraphs and section breaks. I also have the luxury of citing stats.

Canadian Wool by the Numbers

According to StatsCan, in 2023 there were 854,400 sheep on 8,487 farms across the country.[1] Ontario, Québec, and Alberta raise most of the sheep in Canada and there are more than fifty recognized breeds being raised on these farms.[2] The majority of the sheep industry in Canada is focused on meat (lamb) production. In 2022, Canada produced 17,900 tonnes of lamb and mutton and exported $2.4 million worth of live sheep.[3] The reason many knitters don't know that Canadian wool is a thing is because wool is secondary to the lamb/meat industry.

Canada's sheep population is relatively small; it comprises less than 1 percent of the global sheep population and accounts for only 2 percent of the farms in Canada. This hasn't always been the case. At one point Canada had a much larger sheep industry that worked in tandem with the textile industry. This combination had the capacity to grow and process significantly larger amounts of wool than are grown and processed today.[4]

Canadian Wool Goes Out

Statistics Canada (StatCan) reported in 2022 that 1,357,900 kg of wool was purchased from Canadian farms and 48,700 kg was used on-farm. The average price paid to producers for this wool was $0.38/kg.[5] There

---

1 Government of Canada, *Number of sheep and lambs on farms (× 1,000)*, Table: 32-10-0129-01, Statistics Canada, Release date 2025-02-25, https://www150.statcan.gc.ca/t1/tbl1/en/tv.action?pid=3210012901.

2 The Canadian Sheep Breeders Association has a list of over 50 registered purebred sheep in Canada on their website (https://sheepbreeders.ca/home-page). There are many more crossbreed sheep on farms in Canada, and some breeds have historically been found in Canada but may not be on the list now. The list acknowledges only purebred sheep that are registered through Canadian Livestock Records Corporation.

3 Government of Canada, *Market Overview—Canadian lamb/sheep meat and product trends*, Agriculture Canada, https://agriculture.canada.ca/en/international-trade/market-intelligence/reports/customized-report-services-market-overview-canadian-lambsheep-meat-and-product-trends#b.

4 Government of Canada, Dominion Bureau of Statistics Industry and Merchandising Division, "The Wool Textile Industries 1953." Vol.2 – Part VII—D-I, Government of Canada Publications.

5 Government of Canada, *Sheep statistics, quantity of raw wool purchased, used on farms, price and farm value*, Table: 32-10-0142-01, Statistics Canada, Release date: April 19, 2024, https://www150.statcan.gc.ca/t1/tbl1/en/tv.action?pid=3210014201.

The Meat Industry and Its Impact on Wool

When compared to other wool-producing nations like Australia and New Zealand, Canada's wool clip is much dirtier, meaning it is contaminated with vegetable matter (straw, hay, wood chips, other feedstuff). There are a variety of reasons for this, one being the cold Canadian climate which means we must feed hay for up to six months a year—hay is more likely to get stuck in the wool than grass or other forages. Many large commercial sheep farms use feeding systems that are efficient for producing high-quality meat but increase wool contamination. This contamination reduces the overall yield that can be sold. We end up in a vicious cycle where farmers are not paid adequately for their wool, so they focus on increased efficiency for meat production and feed their sheep in a way that increases wool contamination, and because of wool contamination farmers are not paid adequately for their wool. Without the financial incentive to break this cycle, the wool becomes dirtier and reinforces the farmers' belief that wool isn't worth anything and is therefore not worth their effort to improve its cleanliness.[6]

are some inherent challenges with this number. StatCan doesn't explicitly map out who was surveyed, although they do indicate, "…those establishments in the target population for which business information is available on Statistics Canada's Business Register excluding specific farms such as institutional farms, community pastures and farms on First Nations reserves."[7] The description goes on to say, "The smallest

6 In every annual report, Canadian Cooperative Wool Growers states the importance of keeping fleeces clean, and they offer a $0.02–0.08/lb. incentive to sheep producers to sell cleaner wool. However, $0.02 more on $0.38/lb. is not much of an incentive to sheep farmers.

7 Government of Canada, *Biannual Livestock Survey*, Statistics Canada, https://www23.statcan.gc.ca/imdb/p2SV.pl?Function=getSurvey&SDDS=3460, (Accessed October 2024.)

units are ineligible to be selected for the sample. Their contribution is estimated and added to the survey results." As I understand it, this exclusion leaves out the wool that is being grown across the country by small fibre farms for direct marketing to the artisan market, handspinners and knitters. Regardless of the omissions, this data provides us with an overall baseline of the majority of the wool being grown in Canada.

StatCan reports from 2022 indicate that 620,162 kg of wool was exported with a value of just over $1.2 million.[8] The amount of wool exported was higher prior to the COVID-19 pandemic (when the supply chains were disrupted by manufacturing shutdowns). However, annual reports from Canadian Cooperative Wool Growers and a variety of reports from StatCan indicate that Canadian wool exports have been increasing over the last 25 years.

So, if an increasing amount of Canadian wool is being exported, how do we explain the growing and expanding domestic fibre arts industry?[9] Spoiler alert: The majority of the wool you knit/crochet/craft with is imported to Canada from other places.

Other Wool Comes In

Whenever I find a new yarn shop I always ask the staff, "Can you show me your Canadian wool section?"

Often, I am shown a beautiful display of Briggs & Little yarn or Custom Woollen Mill yarn—these two iconic Canadian brands are well represented in many shops. But occasionally the staff person will balk with a confused look on their face and explain that the wool is all from Canada. I will then explain that I'm looking for those brands that use Canadian-grown wool. Usually at this point the staff take me to the

8 Government of Canada, *Customized Report Services—Market Overview—Canadian lamb/sheep meat and product trends*, Agriculture Canada. https://agriculture.canada.ca/en/international-trade/market-intelligence/reports/customized-report-services-market-overview-canadian-lambsheep-meat-and-product-trends#b.

9 Eva Koronios, "Fabric, Craft & Sewing Supplies Stores in Canada Doing Sew Well: Increased Operating Efficiency Among Leading Operators Is Anticipated to Preserve Industry Profit," *Craft Industry Alliance*, May 2020, https://craftindustryalliance.org/wp-content/uploads/2021/11/May-2020-Fabric-Craft-Sewing-Supplies-Stores-in-Canada-Industry-Report.pdf.

hand-dyed Canadian yarn companies where they pick up a skein or two and then exclaim, "Oh this one is made in Canada." I will thank them for their time and turn the skein over to examine the yarn. Usually, it is beautifully hand-dyed in Canada—but the wool was not grown in Canada and the yarn was not spun in a Canadian mill.

According to StatCan, the total retail sales of sewing and knitting supplies in 2021 was over $653 million.[10] We have to dig deeper into these findings to discover that fabric accounts for at least half of these sales, but knitting supplies, including yarn, accounts for 25.8 percent, or more than $168 million in sales (though yarn is categorized alongside other notions in this category and they do not specify what is pure wool or synthetic yarn).

We can focus this picture even more by looking at imports of raw wool in Canada—approximately 605,916 kg in 2022.[11] This is predominantly washed but not carded wool. There is limited information about what this wool is transformed into when it hits our shores, but we have a clearer picture when we look at the imports of "yarn of wool/fine animal hair put up for retails" as it is described by the Canadian Importers Database. Data from 2022 (the most recent data available) reports the value of all wool yarn imports for retail sale as $4.4 million. We can expect that number to double, triple or even quadruple once it is marked up for retail sale.[12] Some of the big name wool yarn importers may be familiar to you: Amtex Yarn Manufacturing, Diamond Yarn of Canada, Mary Maxim, Nordic Yarn Imports, SweetGeorgia Yarns, Artfil Yarn, Berroco Yarn, Hand Maiden Fine Yarn, The Knitting Loft, Estelle Yarns, and Spinrite Yarns. Some of these brands, like Berroco, make and market their own lines of yarn and some are larger distributors of other brands that exclusively sell wholesale to local yarn stores.

10 Ibid.

11 Government of Canada, *Customized Report Services—Market Overview—Canadian lamb/sheep meat and product trends*. Agriculture Canada. https://agriculture.canada.ca/en/international-trade/market-intelligence/reports/customized-report-services-market-overview-canadian-lambsheep-meat-and-product-trends#b.

12 Government of Canada, Canadian Importers Database, https://ised-isde.canada.ca/app/ixb/cid-bdic/productReportHS10.html?hsCode=5109900000&wbdisable=true (Accessed: October 2024).

The commercial wool yarn industry in Canada is a multimillion-dollar industry. Unfortunately, this industry is mainly using imported yarns made with wool from other places and with little or no traceability information readily available. Understanding the "true cost" of wool yarn can be difficult when the supply chain spans vast distances, and it is next to impossible to identify the farm where the wool was grown, the mill where it was processed, and the dye houses that transformed it into yarn. In the following chapters, my goal is to demystify this value chain, equipping all of us with the knowledge and insights needed to make informed decisions about the materials we purchase.

Who Gets to Be a Farmer?

My speaker bio starts with the sentence, "Anna is a first-generation fibre farmer." It's a label and also how I situate myself among the various farming communities. My husband, Luke, and I came into this without access to family-owned farmland, without infrastructure, equipment or farm capital. We also came into it without learned or lived knowledge of how to run a farm. Everything we have done we had to learn from books, YouTube, peers, and mentors.

When I teach or speak about our experience and the "true cost of raising sheep," I often start with the cost of feed, or the cost of purchasing sheep, or even the costs associated with land regeneration—but before any of that comes into play we must recognize that to grow wool and raise sheep, a farm is required. Sheep farms come in all shapes and sizes—some are small with a few sheep on a couple of acres, some have hundreds of sheep on large ranges, and some have many sheep housed in large barns and don't require as much farmland. Regardless of the size and configuration of a farm, land is required. Access to land is becoming increasingly challenging, especially when you consider that the cost of Canadian farmland has increased by 300–900 percent since 1985.[13]

13 Government of Canada, Farm Credit Canada, *Historic FCC Farmland Values*. March 13, 2023, Government of Canada Publications, https://publications.gc.ca/collections/collection_2024/fac-fcc/CC213-7-2023-eng.pdf.

Yes, you read that correctly: Farmland was as much as 900 percent more expensive in 2023 than it was in 1985. (There are several reasons for this, but that discussion is outside the scope of this book.)

If you are raising sheep in Canada there are a few options for accessing land: You can buy land to farm, or you can rent or lease land from someone else. There are some organizations such as the Young Agrarians that work to help build networks for young people looking to access land, but for the most part people who farm sheep either own the land or pay for renting/leasing it. For those who don't own farmland or won't inherit it in the near future, access to affordable farms is decreasing due to the rising expense.

According to Farm Credit Canada, the average value of cultivated farmland increased by 5.4 percent in 2020, 8.3 percent in 2021, and a whopping 12.8 percent in 2022. The average price for a hectare in the Canadian Prairies (where over 70 percent of Canada's agricultural land is located) is between $6,670 and $16,000, depending on location and other factors like access to irrigation.[14]

Access to land for new farmers is increasingly challenging as our agriculture system has shifted away from smaller family farms and local, decentralized food and fibre economies to a system of concentrated control of farmland. In the past 35 years, Canada has seen a significant decrease in the number of small farms (under 1,000 acres). At the same time the number of large farms (over 5,000 acres) has increased.[15] This directly impacts the ability of new or young farmers to access land and begin farming, as many new farmers generally start with small farms.

14 Government of Canada, Farm Credit Canada, *2023 FFC Farmland Values Report*. March 12, 2024, Government of Canada Publications, https://www.fcc-fac.ca/en/knowledge/economics/farmland-values-report.html#:~:text=Farmland%20continued%20steady%20climb,2021%20and%205.4%25%20in%202020.

15 Darrin Qualman, Annette Aurélie Desmarais, André Magnan, and Mengistu Wendimu. "Concentration Matters: Farmland Inequality on the Prairies," *Canadian Centre for Policy Alternatives*, CCPA 2020, https://policyalternatives.ca/sites/default/files/uploads/publications/Saskatchewan%20Office%2C%20Manitoba%20Office/2020/11/Farmland%20Concentration.pdf.

The Canadian Centre for Policy Alternatives summed up this cause-and-effect situation in the 2020 report, "Concentration Matters: Farmland Inequality on the Prairies":

> "Who has the greatest ability to pay for a given acre of farmland? Often it is those who already have the most acres paid for—often those who operate the largest farms. Because margins are tight and per-acre net income is low on cattle farms and grain and oilseed farms, a young or new farmer on a small farm with few acres paid for has a very limited ability to pay for additional acres."[16]

The challenges facing small-scale farmers who want to access farmland and farming capital are further compounded by additional barriers impacting anyone who doesn't fit the traditional image of a farmer when it comes to race, gender, sexual orientation, or newcomer status. Systemic racism and exclusion, and outdated regulations—such as the Indian Act[17]—have historically restricted access to land and financial resources for Black or Indigenous people, people of colour, LGBTQ individuals, and newcomers in Canadian agriculture. Lack of inherited knowledge, intergenerational equity, and local farming expertise further deepens the barriers, leaving new entrants at a significant disadvantage in an industry built on informal, generational knowledge and resource transfer.

Securing land is only the first step to starting a farm; next is the investment in infrastructure. Even small farms have to have fencing, shelter, and transportation for sheep and hay. This of course all comes at a price.

---

16 Ibid

17 "Federal law restricts land use for on-reserve First Nations through the Indian Act. These restrictions, combined with the historical imposition of homesteads, regardless of existing Indigenous communities' use of lands, imposed a regime designed to disconnect First Nations from their traditional territories. It also limited the practices of their agricultural expertise. There are strong correlations between these acts and today's underrepresentation of both First Nations and broader Indigenous communities in farming activities." "Growing the Next Crop of Canadian Farmers," *Action Canada*, https://actioncanada.ca/wp-content/uploads/2021/06/TF-3-WEB-ENG.pdf.

Farming is a capital-intensive industry with high costs for land, equipment, and infrastructure. When we started our farm we went to traditional banks and even to Farm Credit Canada, a financial institution dedicated to entrepreneurs in the agriculture industry, but were repeatedly denied operating loans and lines of credit because we couldn't prove that our farm enterprise would make any money, or we didn't have the collateral to put up against a farm loan.

We decided to approach the problem a little less conventionally and launched a Kickstarter campaign to crowdsource fundraising for our initial farm enterprise. We raised $7,500 from 115 people who believed in our farm business and we were able to build the basic infrastructure for our farm. Following that initial investment from Kickstarter supporters we continued to finance our farm through off-farm jobs and reinvesting any profit we made back into the business. Of course the sort of time, energy, social capital, and fundraising we had access to is not within reach for everyone.

As we consider the true cost of wool in Canada we must acknowledge the start-up costs for establishing a farm in Canada, and the barriers that many new fibre farmers face in accessing land before they have even started growing wool.

GROWING GOOD WOOL

The wool yarn and fibre we use has its beginning well before it arrives at the mill, even before the shearer shows up on the farm.

In a typical season it starts for me on a cool May morning. The sun is just peeking out of the tree line, a chorus of frogs sings as they awaken from their winter slumber. I fumble with the lock to open up the gate, rubbing the sleep out of my eyes. I've been doing "lamb checks" every three hours for a week now and my mid-40s body is definitely feeling it.

I hear the unmistakable sound of a ewe in labour, calling out in the way that is familiar to anyone who has witnessed birth, whether human or animal. I quickly lock the gate behind me and race into the barn just in time to witness the messy and beautiful *sploosh* of a new lamb being born.

The ewe immediately begins making her bleating noises. It is a unique sound for her baby alone, a sound that calls them close and reassures. She starts licking her lamb clean, an important part of maternal bonding, but also to ensure the lamb is dry and won't get hypothermic. As the ewe's pink tongue methodically goes over her baby's mouth and nose, the lamb shakes her tiny head and gives a little sneeze. "Welcome to the world, little one," I say. Birth is magical every time I witness it.

This. This is the start of your yarn.

It was 2019 and I was at Knit City Vancouver and it seemed as though every second person was wearing a *Love Note Sweater*—the beautiful design by Tin Can Knits that looks good on every body. I saw versions of this sweater knit up in soft, subtle colours and in bold, bright, sparkly colours. I saw it in mohair and silk and in natural un-dyed yarn. It was a beautiful display of the variety of yarn available to us, a true celebration of colour and texture and wool.

But what do we really know—as knitters or crafters—about what it takes to grow the wool for our knitting? What do we understand from a chemical, morphological, or environmental perspective about what makes this yarn or that yarn better for a specific project? And what is required of the farmer to ensure that wool is of the highest quality and will perform the way we want it to?

Follicle Facts

Did you know that a sheep has more than 50 million follicles embedded in their skin?[18] These follicles are directly responsible for the wool that we love to work with. (Stick with me—this is going to get a bit wool nerdy!).

There are primary and secondary follicles. Primary follicles are where the sweat glands are and where lanolin is excreted from, it is also

18 W.S. Simpson and G.H. Crawshaw, Wool: Science and Technology. Woodhead Publishing Ltd. 2002.

where more of the coarse fibres grow on a sheep. Secondary follicles are more numerous, there are no sweat glands in these follicles, and they do not secrete lanolin. They are responsible for the finer fibres and the overall density of a sheep's fleece. The secondary follicles are important to us as wool enthusiasts and shepherds alike if we want a lot of good quality wool from our sheep.[19]

A lamb is born with a permanent number of secondary follicles. As fibre farmers, we want a lot of secondary follicles in all of our sheep to ensure a good wool harvest. Secondary follicle development happens in the last two months of gestation (in a five-month pregnancy), so it is vital that the ewe is in excellent health and in good condition during this stage of gestation. She must have adequate nutrition, including protein and minerals, to ensure that extra energy can go toward her fetus and its secondary follicle development.

Wool is a complex protein compound that is formed from the feed that sheep eat. My sheep transform the grasses, legumes, forbs and grain they eat into energy that supports basic maintenance and growth. Leftover energy will be then used to support pregnancy and lactation and finally wool growth. Wool is the final priority for the nutrients a sheep consumes and the animal will sacrifice wool production if there are not enough nutrients supplied.[20]

As sheep farmers we must ensure that our sheep receive the best care and nutrition possible to ensure that our wool is the highest quality. A sheep requires about 11–13 kg of quality feed (50 percent total digestible nutrients) to produce 1 lb. of quality fibre (that is about 4.5 skeins of yarn). Zinc deficiency in sheep can result in brittle fibre that lacks lustre, and a copper deficiency can cause changes in wool colour (whereas too much copper is actually fatal to sheep).[21]

---

19 W.S. Simpson and G.H. Crawshaw, *Wool: Science and Technology*. Woodhead Publishing Ltd. 2002.

20 Eds R. Hergenhan, "Nutritional Determinants of Wool Quality and Quantity," The Australian Wool Trust, University of New England, 2014, https://www.woolwise.com/wp-content/uploads/2017/07/WOOL-300-300-14-T-04.pdf.

21 W.S. Simpson and G.H. Crawshaw, *Wool: Science and Technology*. Woodhead Publishing Ltd. 2002.

So, What Does It Actually Cost to Grow Quality Wool?

The day-to-day costs of running a sheep farm vary depending on the farm's overall management style, how many sheep they have, and what values are brought into their farming operation. Obviously, a book like this cannot adequately cover the entire range of costs for every different farm out there. Instead I will offer some insight into the costs associated with farming from a more general view. This will hopefully give those that are not farming a look into the true cost of raising sheep. And to those who are raising sheep and find these costs either too high or too low, I encourage you to share your own true costs with your customers. We all have different contexts that influence our own costs and pricing. This is also a reminder that with such a large land base here in Canada, the situation for sheep farmers on the Prairies varies greatly from those raising sheep on, say, the Gulf Islands of British Columbia. Farmers dealing with drought will experience vastly higher feed prices than those who had a year with adequate precipitation. The following figures are based on my own experience farming a flock of Shetland sheep and the various interviews I have done with other farmers across Canada, along with industry statistics and reporting. I have tried to provide a few different scenarios to explain the variation in production costs.

FOLLOW THE WOOL

Note: You may want to bookmark this page so you can easily come back to these three scenarios, which we'll revisit throughout the book.

Scenario 1

This scenario involves the cost of raising one sheep for one full year for a fibre farmer who is focused solely on wool production in a pasture-based operation. In this scenario the farmer has a small flock of forty sheep that are raised exclusively for wool and agritourism (they are not selling meat and don't rely on breeding stock sales).

They direct-market the wool, meaning they sell value-added wool products direct to customers. They do have other contract work, so the fixed costs are based on a percentage of the sheep operation. The cost of production was calculated by taking the total of all costs throughout one year and then dividing by the number of sheep in the operation (for a more detailed explanation see Appendix).

Cost of Production for 1 Sheep for 1 Year

Fixed Costs: $54.65

Variable Costs: $115.65

TOTAL: $170.30

- Amount of wool from 1 sheep in 1 year: 5 lb. coloured wool[22]
- Cost of growing wool per 100 g skein of yarn: $15.07
- This would produce approximately 11 skeins of 100 g yarn.

How did I get this number?

- 5 lb. of greasy wool will lose approximately 40–50% of its weight during washing and processing. I am choosing to stick with 50% loss for all calculations to be most accurate/conservative.
- That leaves us with 2.5 lb. of finished wool yarn product.
- This is 1133 g or 11.3 skeins of yarn, each 100 g. I took the total cost of raising one sheep for a year—$170.30—and divided it by 11.3 for a total per-skein cost of $15.07.

Scenario 2

Some sheep farmers in Canada focus on wool production but also sell lamb meat or breeding stock. This means that their revenue stream is diversified and they do not rely exclusively on the sale of wool to cover their costs. Here is a scenario of a sheep producer

22 Numbers and figures from Long Way Homestead sheep farm, December 2023.

that is focusing on wool and lamb/breeding stock production. In this example I looked at a flock of 100 Rambouillet sheep in Western Manitoba. I took the cost of production for the entire flock and then divided it by the number of sheep to get the individual cost for 1 sheep.[23]

Cost of Production for 1 Sheep for 1 Year
Fixed Costs: $83.00
Variable Costs: $167.99
TOTAL: $250.99

- Amount of wool from 1 sheep in 1 year: 8.5 lb. fine wool
- Cost of growing wool per 100 g skein of yarn: $13.07
- 8.5 lb. at 50% loss = 4.25 lb./1929.5 g, or 19.2 skeins at 100 g each
- $250.99/19.2 = $13.07/100 g skein

Scenario 3

This final scenario represents the majority of Canadian sheep farmers. They focus on producing lamb and breeding stock for the commercial food industry, and wool is a by-product of their operation. Some of these farmers see wool as a valuable product and work to keep it as clean as possible, while others consider wool as a way to keep sheep warm and view shearing as a "cost of doing business."

This is the price per sheep in a 225-head, intensively managed, indoor operation.[24]

23 Numbers shared by Graham Rannie of GRannie Rambouillets in Russel, MB, 2023.

24 2022 Cost of Lamb Production, Ursina Studhalter, Jan 28, 2023, Sources: website and email interview.

Cost of Production for 1 Sheep for 1 Year

Fixed Costs: $48.31

Variable Costs: $117.22

TOTAL: $165.53

- Total amount of wool from 1 sheep in 1 year: 8 lb. medium wool
  Cost of growing wool per 100 g skein of yarn: $8.80
- 8 lb. at 50% loss = 4 lb./1816 g, or 18.6 skeins at 100 g each
- $165.53/18.6 = $8.80/100 g skein

That is a lot of numbers, and if you aren't involved in the growing and buying/selling of wool it might be challenging to understand what it even means. So I will break it down further, excluding the cost of buying and setting up a sheep farming operation and distilling the numbers so we consider what it costs to grow 100 g of wool, the fibre content of one skein of wool yarn (this of course is looking only at the cost of growing the wool; it does not include all the work to make it into yarn).

Summary of Costs to Grow 100 g of Wool

***Scenario 1:* $15.07/100 g skein**
***Scenario 2:* $13.07/100 g skein**
***Scenario 3:* $8.80/100 g skein**

SHEARING

I've watched about 5,000 sheep be sheared in the years since I became involved in the agricultural end of wool production. This number pales in comparison to the number of sheep that shearers have actually sheared, but is probably more than the average knitter has seen. Watching shearers never fails to fascinate me—I have always been amazed by the calm, cool manner of every single shearer I have ever met. Eventually, I imagined that I had witnessed enough shearing that I could probably do it (in a pinch) without too much struggle.

I could not have been more wrong. My over-confidence has been my greatest strength and my greatest weakness. In this situation it proved to be pure hubris.

In 2024, I attended the shearing day at my good friend's farm. He raises Rambouillet and Border Cheviot sheep in Russel, Manitoba, and had hired some expert shearers for the two-day shearing event—the shearers had just returned from a six-month shearing stint in Australia. When the shearers offered to help me through shearing my first sheep I nervously said sure. Let me be clear—I was nervous, but I also thought I could do it, no problem.

It was horrific. Don't worry, the sheep was totally unscathed. But I was dripping in sweat, my hair plastered to my face by sweat, lanolin and manure. I felt like I wouldn't be able to stand up straight for the next week without something hurting—and it took me twenty minutes to shear two-thirds of the sheep. Let's be honest—I probably should've paid the farmer twice for that fleece, because I'm pretty sure the shearer had to go back over all the sections I did, to clean it up.

Shearing is hard work, and every single skein of wool yarn that you have ever worked with was sheared by hand—by an expert who can easily shear a sheep in under two-and-a-half minutes!

Since humans domesticated sheep more than 10,000 years ago, and bred out the rooing instinct (the natural shedding of wool) we have had to shear sheep. Shearing sheep involves a coordinated and rehearsed choreography that ensures the safety of both shearer and sheep, a dance that when done correctly is efficient and results in a quality fleece for us to use to its maximum potential.

We can thank Godfrey Bowen, and his revolutionary changes to the way sheep are sheared, for the quality of the wool we enjoy today. Bowen was originally from New Zealand and in 1953 he broke a world record by shearing 456 full-wool ewes in nine hours. This result, and his continued work to promote shearing as a spectator and competitive sport, catapulted him to the chief shearing instructor position at the New Zealand Wool Board. The Bowen method involves using the non-shearing hand to stretch out the skin on the sheep ahead of the hand-piece blade, resulting in a more evenly shorn fleece. The outcome of this technique is that less wool is wasted and the fleece is of higher value to the shepherd. This method became popular all over the world, securing Godfrey Bowen's impact on modern-day shearing and competitions like the Golden Shears.[25]

There is a sheep-shearing industry in every country with sheep. Some shearers do it full time and some shear competitively, while some have the skills just to shear their own flock once a year. And despite the promise from Australian Wool Innovation of biological wool harvesting or defleecing (a process being tested in Australia where sheep are given an injection of a biological agent to create a break in the fleece so that it can be removed by machine without skilled labour),[26] we still need human shearers.

In Canada, there are currently ninety-six shearers listed on the Canadian Cooperative Wool Growers (CCWG) website.[27] Of course there are many more in each region across the country that aren't listed, and many farmers shear their own sheep. But 96 shearers to shear more than 854,400 sheep is roughly 8,000 sheep per shearer per year. My body aches just considering that workload.

Canada is a large country with sheep farming in every province. Some farms have huge flocks and some have only a handful of sheep. There are many different breeds represented—some farms even have more than five breeds within their flock. All of this requires skill and flexibility from a sheep shearer. This becomes extra complicated as the weather is often unpredictable and many farms do not operate a climate-controlled environment such as a barn, and have to pivot as rain or snow can change the timing of a shearing event.

25 Government of New Zealand, Manatū Taonga—Ministry for Culture and Heritage, Godfrey Bowen, https://nzhistory.govt.nz/people/godfrey-bowen, updated 8-Nov-2017.

26 Australian Wool Innovation Ltd., https://www.wool.com/news-events/news/biological-harvesting/ (Accessed October 2024).

27 For a full list of shearers across Canada, check out the Canadian Cooperative Wool Growers website: https://wool.ca/page/shearers.

A shearing friend of mine once said she feels like she spends more time driving to shear sheep than actually shearing sheep. Producers have to cover the shearer's mileage fee and often are at the whim of the shearer's schedule—if the shearer actually makes it to their farm at all (especially if their farm is off the beaten path). A shearer travels with their own equipment but the farmer is required to have the sheep penned up, out of the weather. It is dangerous for both sheep and shearer to shear wet sheep.

Shearing fees can vary by region, and because shearing is not standardized and there is no national shearing organization, pricing is up to the individual shearer. I've heard of shearing costing between $5.00 and $15.00 per head (depending on the conditions), plus mileage. The mileage fee isn't that exorbitant if you have a large flock and the fee is spread out among all of the sheep, or if the farmer lives close to a shearer, or to other farms that also need a shearer. But it can be exceptionally pricey for those who are farther afield.

FOLLOW THE WOOL

Cost of Shearing 1 Sheep

Here are the shearing costs for our three scenarios. The difference in the costs are due to mileage, the total number of sheep, and the individual rates of shearers across the country.

Scenario 1

Shearing: $5.00
Mileage: $1.80
TOTAL: $6.80

Scenario 2

Shearing: $5.00
Mileage: $2.00
TOTAL: $7.00

Scenario 3
Shearing: $6.00
Mileage: $1.00
TOTAL: $7.00

We can further extrapolate this shearing cost to a single skein in each of our three scenarios by taking the total cost of shearing one sheep and dividing it by the total number of 100 g skeins of yarn that sheep will produce.

Summary of Shearing Costs

***Scenario 1:* $0.60/100 g skein**
***Scenario 2:* $0.36/100 g skein**
***Scenario 3:* $0.37/100 g skein**

Most sheep shearers in Canada have taken charge of their own training, as there is no standard education. On our farm, we've had a shearer who learned from a book, and a shearer who learned completely from YouTube. I've attended shearing and wool-handling training sessions where professional shearers were brought in to train new and seasoned shearers, and I know that mentorship is often how new (especially young) shearers break into the business. Shearing takes hours of dedicated practice and is a skill that cannot be taken for granted. For sheep farmers across the country, the importance of having access to timely and affordable shearing cannot be understated—without it we would not have any wool at all. Every skein of yarn you have ever used or worn was sheared by hand.

After shearing, the path that the wool travels is almost entirely up to the producer. Dedicated fibre farmers who are growing wool specifically for wool production or sale in the fibre arts industry will skirt and then grade their fleeces for sale. Skirting is the process of pulling off the dirty or contaminated parts of a fleece after shearing. Grading a fleece involves evaluating the fleece for quality, soundness, colour, staple length, and condition. Fibre farmers store their wool according to animal, breed, colour, or other factors that are unique to their customers. They will then transport their wool to a mill for processing or start marketing the raw fleece.

FOLLOW THE WOOL

Direct to Consumer

The responsibility for processing and selling wool falls entirely on the shoulders of the producer that grew it. On one hand, this ensures enhanced traceability, because the producer engages with every stage of the value chain. On the other hand, it requires more work for the producer, because they have to be involved in every stage to ensure the highest return on their wool. Most fibre producers in Canada choose one of the following options for marketing their wool:

- Sell large batches to a local mill or a small-scale wool broker
- Have their wool processed at a mill into value-added products (yarn or spinning fibre) and sell to fibre artists directly
- Manufacture a unique value-added product—like blankets, bedding, insulation, etc.—and sell it direct to consumer or wholesale to resellers

There are many innovative sheep farmers in Canada who have worked hard to navigate each step of the process that comes after the wool is grown, and have, in most cases, developed their own market for their products. These producers stand out as creative entrepreneurs and leaders in the industry. They possess a unique set of skills to not only grow the wool, but to be the main seller as well. In the next chapter we will uncover the true cost of wool processing in Canada for those who choose this path.

Commercial Wool

Commercial sheep producers have a few options on shearing day. Shearing is always fast-paced and busy, especially when you are dealing with 500 to 1,000 or more sheep. The extra work of skirting and sorting wool is not prioritized if it doesn't make financial sense for the producer. Why put in extra effort if it won't result in significant increase in per-kilogram pay rate? So the next steps a producer takes will depend on their end goal for the wool, and on whether they consider the wool as a resource or not.[28] They usually pursue one of the following paths:

- Wool has no value and enters the waste stream (compost, garbage, burn pile),
- Wool has minimal value and is shipped to CCWG without being skirted or sorted,
- Wool has different value based on its quality and will be sorted and sold based on that value.

Because most of the wool grown in Canada is grown on commercial farms, the majority of our wool clip falls into one of the three options described in the previous Follow the Wool segment, and this is the part of the story that has motivated me the most to make a change in the industry. In 2016, when I first started meeting with producers and going to shearing days in larger sheep barns, I was shocked at what I discovered.

Many farmers, disillusioned by the insignificant value of wool and the meagre size of their wool cheque every year, had decided that the cost and effort of bagging and shipping wool wasn't worth it, and were simply putting their wool into various waste streams. Many producers

28 I can't quantify the number of sheep producers I have spoken with who have echoed these sentiments. It's too many for me to remember, but enough that I didn't have to think twice about these three salient points. These three options are not outlined specifically in any sort of sheep handbook or resource guide, but anecdotally it is what happens to the majority of the Canadian wool clip.

would burn their yearly wool clip, bury it in the ground, or take it to the local landfill.

As a new-to-the-industry business owner and an idealist through-and-through, I was appalled. I was convinced that I could change this practice. Since then, I've spent more time with sheep farmers and come to understand what a big hill we have to climb if we actually want the majority of Canadian wool to be valued, and for producers to take the time to prevent wool contamination and see the opportunity in their wool clip. The responsibility is on both producer and industry to change the mindset around the value of wool, to change management practices to prevent contamination and improve wool quality, and to increase the value of wool to incentivize the extra effort and work in management.

Canadian sheep farmers will benefit greatly if we can revalue all the raw materials that come from their flocks. Imagine the opportunity and long-term security for farmers that would result if they could generate revenue not only from lamb and breeding stock but from wool as well. I have a few ideas about how we can get to that point, that I will share in the last chapter of the book, but for the moment let's attempt to follow commercial wool into the production pipeline and unearth some of the true costs that the majority of our wool clip reveals.

The next step for commercial producers and fibre farmers is shipping and freight. For those producers that don't trash their wool clip, they will find the closest CCWG depot and arrange for transport of their wool.

CHAPTER 2

Shipping Wool

Running a small-scale wool operation in Canada, I think I have moved wool by every means possible (and some have been downright sketchy).

I once transported twenty-two garbage bags of fleece, each weighing eight to ten pounds, in my 2003 Toyota Yaris—with my kids sandwiched in between the bags in the back seat. During the trip, my six-year-old said, "At least if we get in an accident we will have lots of soft things to land on." Perhaps not my finest parenting moment.

Another time, I was twenty-five pounds under the max load of my rental pickup truck when I loaded five-and-a-half burlap wool grower bags, each weighing 200–300 pounds, into the back. It was a flawless plan until we hit Moose Jaw, Saskatchewan, and it started to pour rain. I pulled over to tarp the load but my tarps were too small. My jeans were soaked, my fingers frozen, hair plastered to my head, and every second vehicle honked at what I'm sure was my butt crack hanging out for all to see.

I once rented a mini van to transport fleeces that happened to have mostly been shorn from rams. If you've never been around a ram, they are extra stinky. I had to provide the rental company with photographic evidence that I hadn't violated the "no animals in the vehicle" clause of my rental agreement before I got the damage deposit back.

With all the money I've spent on rental trucks, I've often wondered why I haven't invested in a livestock trailer—or a larger vehicle for moving wool. But it just doesn't make financial sense to invest in that type of transport for the seasonal shipping of wool, and even if we could justify the expense, we did not have the capital to invest in a trailer. So the next option I turned to was transport trucks.

Navigating the world of transport trucking is something I never thought I would have to do. I initially thought I could just call a transport company and get a quote. But it turns out, attaching wool bags to standard shipping pallets is challenging if the wool bags are the older burlap version. Picture sausage-shaped bag of wool that is seven feet tall and four feet around. Packed well it's quite stiff and will mostly stand up on its own, but these bags are often a bit floppy, making it impossible to load as a regular transport item. It's a big waste of time to repack 300 pounds of wool into a newer, nylon standard wool shipping bag just so you can load it on a pallet without it falling over.

So, at some point, it was suggested that I find a trucker who would be willing to "backhaul" wool bags for me under the table. I first started working with...let's call him Dave[1]...to get my wool across the country. We would meet up at the closest truck stop in Richer, Manitoba—a place called the Cat Sass (I'm not joking with that name; they have great fries, but skip the chocolate cake).

Dave would backhaul wool for me to and from the scouring mill in Ontario and I would pay him in cash between $200 and $300 per trip depending on how much wool I was sending. Because we would meet up in a parking lot, my husband and I had to physically move the wool from the transport trailer to our truck bed. It was easy when it was a 250-pound bag of tightly packed raw wool—less easy when it was a five-foot-high bale of scoured clean wool weighing 650 pounds and wrapped in slippery plastic (the scouring mill doesn't want to pack clean wool into the dirty wool bags, so they transport them in thick plastic bags).

Three years in, we found a trucking company that is more than happy to transport our wool reliably and fairly affordably. As our business has grown, I have stopped transporting wool in garbage bags in my car or in poorly filled wool bags, and shifted to a more efficient and cost-effective process. This new process has made me consider the

1 Name changed.

implications and necessity of how and why we transport wool in Canada. Canada is 7,821 kilometres from the Pacific coast to the Atlantic coast. Wool is heavy to transport—more often than not it is transported unwashed, which means it includes the weight of the lanolin and dirt, which can be up to 50 percent of the entire weight of the wool.

On many commercial farms, wool is packed with a hydraulic press into either a long jute sausage-like bag or a square polyethylene bag. The bags can hold from 100 to 350 pounds of packed wool and measure approximately 40 × 48 × 40 inches. They are heavy, solid bricks of wool that need to be moved to the appropriate destination. How we move them has a cost—both an obvious price and a hidden environmental cost.

FIBRE FARMING FREIGHT

Sheep farmers in Canada have to decide how much they are willing to pay to move their wool, and what mode of transportation they will use. If they are selling it to the CCWG they will most often load up a truck, trailer, or livestock trailer with their bags of wool and drive to their local wool depot. However, if a farmer is hoping to have their wool processed domestically, they are responsible for getting their wool to a mill for processing.

Shipping wool has some inherent challenges:

- Wool is almost twice as heavy before it is scoured (washed), because of the dirt and lanolin it contains. So farmers are literally paying to ship dirt.
- Canada is huge, and wool often has to travel many hundreds of kilometres to get to the closest mill.

With fewer than fifty mills across the country for processing wool, many sheep farmers spend hundreds of dollars each year to transport their wool to a mill. Here are some examples that include the transport costs.

FOLLOW THE WOOL

Wool Transport from Fibre Farms

Scenario 1

A small rural fibre farmer who must use Canada Post to ship their wool to a mill that is approximately 500 km away (Canada Post is the only option in their rural community). The farm has 25 sheep, each yielding about 5 lb. of wool.[2] Wool loses about 50% of its weight during processing, so 125 lb. will end up as 62.5 lb. for return transport, and in the end will produce approximately 283.75 skeins of yarn.

- Amount of wool to transport: 125 lb.
- Method of transport: Canada Post
- Freight cost to mill: $150.82
- Skeins of yarn: 283.75
- Return freight to farmer: $73.49

Total freight cost: $224.31
Transport cost/skeins of yarn: $224.31/283.75
= $0.79 freight cost per skein

Scenario 2

A dual-purpose (meat and wool) flock. The farmer processes 200 lb. of fine Rambouillet wool at a local mill about 780 km round trip from their farm, and transports the wool using their own vehicle. This involves two trips—one to drop off the raw wool and one to pick up the processed wool. Wool loses about 50% of its weight during processing, so 200 lb. will end up as 100 lb. for return transport and will produce approximately 454 skeins of yarn.

2 Shipping estimates were based on the online rate calculator at www.canadapost.ca.

- Amount of wool to transport: 200 lb.
- Method of transport: Personal truck
- Freight cost to mill: $233.90
- Skeins of yarn: 454
- Return freight to farmer: $233.90[3]

Total Freight: $467.80
Transport cost/skeins of yarn: $467.80/454
= $1.03 freight cost per skein

Scenario 3

A lamb- and meat-production flock that will process 300 lb. of medium wool at a mill only 178 km away, round trip. Some producers have access to nearby processing and can reduce their transport costs accordingly. Wool loses about 50% of its weight during processing, so that 300 lb. will end up as 150 lb. for return transport and will produce approximately 681 skeins of yarn.

- Amount of wool to transport: 300 lb.
- Method of transport: Personal truck
- Freight cost to mill: $115.88
- Skeins of yarn: 681
- Return freight to farmer: $115.88[4]

Total Freight: $231.76
Transport cost/skeins of yarn: $115.88/681
= $0.34 freight cost per skein

3 This calculation was made using vehicle make and fuel efficiency. I used an online cost calculator with gas costing $1.55/litre in March 2025.

4 This calculation was made using vehicle make and fuel efficiency. I used an online cost calculator with gas costing $1.55/litre in March 2025.

Summary of Shipping Costs
***Scenario 1:* $0.79/100 g skein**
***Scenario 2:* $1.03/100 g skein**
***Scenario 3:* $0.34/100 g skein**

The environmental or emissions costs of this sort of ground transport are worth considering for small-scale wool production, but it is challenging to calculate a general amount of greenhouse gas (GHG) emissions on small-scale fibre transport when there are so many variables involved, including distance travelled, fuel efficiency of vehicles, proximity to processing, and frequency of transport. I have attempted to calculate the GHG emissions associated with our three scenarios to provide a comparison and a starting point as we look at the true cost.

I used online carbon footprint calculators, together with vehicle and fuel efficiency information and the distance travelled, to estimate the emissions measured in carbon dioxide equivalents (CO2e). Canada Post uses carbon-neutral shipping strategies and provide their own carbon dioxide equivalent estimates with their shipping quotes.

Scenario 1 produces **0.004 tonnes CO2e** This is equivalent to 0.0009 passenger cars driven for 1 year. (At the time of this writing, the U.S. Environmental Protection Agency online calculator compares carbon-equivalent emissions to everyday activities. It assumes that one passenger car is being driven for one full year or 17,569 km.)[5]

Scenario 2 produces **0.68 tonnes CO2e.** This is equivalent to 0.140 passenger cars driven for 1 year.

Scenario 3 produces **0.16 tonnes CO2e.** This is equivalent to 0.034 passenger cars driven for one year.

It is important to look at the entire life cycle of our wool, yarn, or garments to understand the full picture of transport emissions. The shipping of wool to and from the mill will produce emissions, but the

5 Or 17,569 km per vehicle per year. For more explanation of this calculation by the EPA: https://www.epa.gov/energy/greenhouse-gas-equivalencies-calculator-calculations-and-references#vehicles. As of March 2025, the link for the EPA's Greenhouse Gas Equivalencies Calculator is https://www.epa.gov/energy/greenhouse-gas-equivalencies-calculator.

local marketing and sale of the finished items means very little emissions on the other end. When businesses and farms that sell wool and wool products have a transparent and traceable supply chain, we can assess the full life cycle of the products in that chain and the full impact those products have on the environment and the economy. We may learn that our yarn travels only a couple hundred kilometres before it lands in our stash, or maybe even less than 100 km if you live in an area with a vibrant regional fibre economy and adequate local processing.

Compared to what we know about yarn that is processed domestically, we can never really know the full picture of wool that is grown, manufactured, dyed, and shipped internationally. There are no labelling requirements for wool products, so the consumer has no insight into the transportation methods or miles that the wool has travelled throughout the supply chain. Without this baseline information we cannot even estimate the emissions generated from the movement of our wool. Given the need to lower emissions globally, this information is important to have. With more transparent labelling and readily available traceability information, consumers can make more informed choices about the transport emissions associated with the products they buy. And with more domestic processing we can limit transport-based emissions.

COMMERCIAL WOOL FREIGHT

Understanding the environmental costs associated with the ways our wool is transported is necessary to understanding the "true cost" of wool. In 2022 the transportation sector was the second-largest contributor of GHG emissions in Canada, accounting for 22 percent of emissions.[6]

Within the transportation sector, 37 percent of all emissions are from transport trucks.[7] There are no specific statistics on how much

6 Government of Canada, https://www.canada.ca/en/environment-climate-change/services/environmental-indicators/greenhouse-gas-emissions.html#transport (Accessed September 2024).

7 Bora Plumptre, Eli Angen, and Diane Zimmerman, "The State of Freight." *Pembina Institute*, 2017, www.pembina.org/pub/state-freight.

of that domestic transport is wool, specifically, but we can extrapolate some numbers from the information we do have about the path Canadian wool takes in the larger commercial industry.

After shearing, farmers will identify the closest CCWG depot in their region and deliver their wool to it. Wool is then transported via transport trucks or rail to the CCWG depot in Carleton Place, Ontario. In their large warehouse, trained CCWG staff sort each bag of wool, then grade and class it according to quality, micron count, colour, staple length, and cleanliness. Wool is then packed in large 1,200-pound bales and waits to be sold to domestic or international buyers. The CCWG attends trade shows and networks with large international wool buyers to find bulk pricing for Canada's wool clip based on how each bale is graded. Once the wool has been sold, it is then shipped to its destination.

Despite repeated requests to the CCWG for accurate and detailed information about the movement of the wool they buy and sell on behalf of Canadian sheep farming shareholders, they were not overly forthcoming in their responses. They would not confirm if they still use rail as a method of shipment for Canadian wool, though they did say that they often use transport truck backhaul as a way to be economical.[8] They would not confirm which Canadian port(s) they use to ship wool overseas. I have done my best to consider various scenarios for how the CCWG could move wool.

My assumption, based on my understanding of international shipping and transport, is that wool purchased within Canada is moved by transport truck to its destination. If sold internationally it could be transported by truck to the Port of Montréal (the closest port to CCWG headquarters in Carleton Place) and from there, shipped by boat to

8 A backhaul load is when a transport truck takes a load back to its origin point on a return trip. Some suggest this is a more affordable option because the transport driver has to return anyway, and it's better than transporting an empty trailer.

international buyers, most of which are Chinese brokers.[9] According to the CCWG 2018 Annual Report, a contract was negotiated to sell 700,000 pounds (317 metric tonnes) of wool to Chinese buyers.[10] I have not been able to source more recent numbers from the CCWG on how much wool has been bought or sold since, so I will use the 2018 numbers.

There is a financial cost to sheep farmers who transport their wool to CCWG wool depots. This cost will vary widely depending on how much wool they grow, how they transport it, and how far the wool depot is from their farm (which is often not terribly far). Once the wool reaches the depot there is no further cost to the farmer, as this is the point of sale. While I would like to calculate the cost to commercial farms of shipping their wool for processing, I have to recognize that because we have no idea what happens to Canadian wool once it enters the international market, I cannot make an accurate estimate of how much it costs per one skein of yarn. I have tried to be consistent with my comparisons and true cost analyses in this book, but the lack of information about what happens to our commodity wool, specifically, and the huge spectrum of what actual shipping costs might be for Canadian commercial farmers makes this number almost impossible to generate. Instead, I will focus on the true environmental cost of commercially grown wool that I can estimate with available information.

By using a carbon calculator we can estimate how much GHG is emitted in the sale and transport of this wool within Canada and then abroad.[11] Employing a bit of citizen science here, we begin to see what the true costs of transporting our wool supply looks like.

9 The CCWG would not confirm how much wool was sold to Chinese buyers, but I did read every Market Update from the General Manager since 2005, and as many annual reports as are available to the public. They repeatedly reference selling wool to Chinese buyers and attending wool trade fairs in China.

10 "Annual Report 2018," *Canadian Cooperative Wool Growers*, February 2018, https://www.wool.ca/images/uploads/files/2018-Annual-Report.pdf.

11 There are many carbon calculators available. I used this one, by CN rail, which focuses specifically on transport-related emissions: https://www.cn.ca/en/delivering-responsibly/environment/emissions/carbon-calculator.

FOLLOW THE WOOL

Wool Transport from Commercial Farms

Scenario A

This scenario involves 290 tonnes of wool going from Alberta farms to a Chinese processing facility by transport truck and marine shipping.

In 2018 the province of Alberta sold 290 tonnes of wool.[12]

- Transportation by transport truck from Alberta to CCWG depot at Carleton Place, Ontario: 76 tonnes of CO2e (carbon dioxide equivalent) emissions
- Transportation by transport truck from Carleton Place to Port of Montréal: 6 tonnes of CO2e
- Transportation by boat from Port of Montréal to Port of Shanghai: 52 tonnes of CO2e

Total CO2e emissions: 134 tonnes

(This is equivalent to 104 passenger cars driven for 1 year[13], or 1,138 barrels of oil consumed.[14])

Scenario B

This scenario involves 290 tonnes of wool going from Alberta farms to a Chinese processing facility by rail and marine shipping.

---

12 Government of Canada, *Sheep statistics, quantity of raw wool purchased, used on farms, price and farm value*, Table: 32-10-0142-01, Statistics Canada, https://www150.statcan.gc.ca/t1/tbl1/en/tv.action?pid=3210014201, Release date April 19, 2024.

13 Or 17,569 km per vehicle per year. For more explanation of this calculation by the EPA check out: https://www.epa.gov/energy/greenhouse-gas-equivalencies-calculator-calculations-and-references#vehicles.

14 All emissions equivalencies were calculated using the online tool from the United States Environmental Protection Agency (EPA), https://19january2017snapshot.epa.gov/energy/greenhouse-gas-equivalencies-calculator_.html.

- Transportation by rail from Alberta to CCWG depot at Carleton Place, Ontario: 13 tonnes of CO2e
- Transportation by rail from Carleton Place to Port of Montréal: 1 tonne of CO2e
- Transportation by boat from Port of Montréal to Port of Shanghai: 52 tonnes of CO2e

Total CO2e emissions: 66 tonnes
(This is equivalent to 51.1 passenger cars driven for 1 year, or 560 barrels of oil consumed.[15])

What is most surprising to me about these two scenarios is how much more carbon-equivalent emissions come from transporting wool by truck compared to rail.

If we want to consider the true cost of moving wool from farm to processing to consumer, we need accurate information and reliable traceability systems. When we have the full picture of the route our wool follows, we can generate a more useful report on the GHG emissions involved so we can make better informed decisions about the products we purchase and use.

As we will learn in the next chapter, a major reason why wool is shipped offshore is the lack of adequate processing infrastructure in Canada. The reasons behind the decimation of our domestic wool-processing infrastructure are complex and will be explored in later chapters. Our raw wool resources are shipped overseas to utilize the labour and infrastructure in low-wage countries. While this results in lower manufacturing costs, it is a system that does not account for the true environmental cost of trans-oceanic shipping or transport-truck emissions.

15 Ibid.

If we want to build a wool supply chain that accounts for and works to limit the GHG emissions of transport, and the burden of shipping on wool producers, then reducing the number of miles wool travels will have the greatest impact. We can do that by building up regional processing and manufacturing facilities that accommodate the quality and quantity of the wool grown in those regions.

Increased cooperation among sheep farmers and consumer organizations or cooperatives to pool wool from a specific region and transport it to local mills could decrease the time and money involved in—and the GHG emissions from—several smaller transport trips. This sort of cooperation and coordination takes communication and time that unfortunately many farmers do not have in abundance. Encouraging a new form of bulk wool buying, transporting, and domestic processing could result in increased profit for the farmers, new products for the domestic market, and less negative environmental impact.

My vision of the Canadian wool industry is one in which local wool is accessible to all, and consumers and producers work together to reduce the financial burden of shipping from the farmer or producer.

I believe that the growth of our domestic wool industry must be decoupled from the GHG emissions of international and domestic shipping and that we should move our industry away from centralized wool collection and processing to a more vibrant and regional system.

CHAPTER 3

Processing Wool in Canada

"Processing mills are the most significant infrastructure investment our communities need in order to decentralize the textile industry and reinvigorate regional economies."

—Rebecca Burgess[1]

Before I started growing wool, I had never really considered the existence, or lack thereof, of wool processing mills in Canada. Despite being an avid knitter for many years, I just sort of took for granted that yarn would be available for purchase at the store. In the early days of my knitting experience, I lived in Halifax, and the only place I knew of to buy wool was Lewiscraft—an all-purpose craft supply store that also sold yarn. Michaels wasn't even a thing in Canada yet. I remember first stumbling across a dedicated yarn store in the Hydrostone Market in Halifax, Nova Scotia. I was blown away—and super excited to find so many yarn options. I bought a black skein of 100 percent wool yarn. It was the first time I had ever bought wool in any form other than a ball and I had no idea it needed to be wound into a ball before I could use it—

1 Rebecca Burgess, *Fibershed: Growing a Movement of Farmers, Fashion Activists, and Makers for a New Textile Economy*, Chelsea Green Publishing, 2019.

it ended up as a total mess, but I was determined to figure out the yarn barf and work with it. I spent hours handling and untangling that yarn, and it still never occurred to me how it was made.

Then I opened up a yarn shop. I had planned to exclusively stock Canadian-grown yarn, and I knocked on every door of every sheep farmer I could find to sell me their wool, but to no avail. Even then, I still didn't ask the question about the manufacturing of yarn.

It wasn't until I moved to Manitoba, purchased my very first sheep and kept them alive for a full year—and then found a shearer to come and shear off their wool—that I realized I had to figure out what to do with all of it. I googled "wool mill in Manitoba" and to my shock there was no mill to be found anywhere in the province. This was the first time I started asking, "What happened to all the mills?" and I've been trying to answer that question for nearly a decade.

The lack of infrastructure to process the Canadian wool clip into valuable and marketable products is, in my opinion, the largest barrier to growing our industry. At one point in our history we processed more than 80 percent of all the wool grown in Canada, which at its peak was close to 11 million pounds. Now we process less than 10 percent of the 3 million pounds grown domestically every year.[2] Of course I don't advocate that we go back to the exploitative factory system of wool and textile production that was typical during the nineteenth and early twentieth centuries. But I also don't believe that we are destined to maintain the status quo system of exporting raw materials to be transformed into products in low-wage countries then importing finished products for retail.

I approach this book, my research, and my long-term work with the belief that we need to create shorter supply chains, not longer ones, and that we can and we must relocalize our wool and textile manufacturing in Canada. When we invest in local value chains we build economic

2 Government of Canada, *Report of the Royal Commission on the Textile Industry*, Justice W. F. A. Turgeon, 1938, Government of Canada Publications, https://epe.lac-bac.gc.ca/100/200/301/pco-bcp/commissions-ef/turgeon1938b-eng/turgeon1938b-part1-eng.pdf.

resilience in our regions and local accountability for that manufacturing. We keep knowledge, wealth, and jobs in our communities and we generate a smaller carbon footprint.

It is important to understand where our industry has come from so we know how to avoid mistakes of the past and how to build on our strengths. My research of the history of our wool textile and yarn industry is not exhaustive, and much of the story is told by the individual mills and companies that comprised the industry throughout the twentieth century. This topic certainly could fill an entire book itself. I aim to provide some historical context and understanding of how the industry has changed in the past 150 years and outline some of the key lessons we need to learn from the past.

There are three main reasons why the Canadian wool manufacturing industry is almost non-existent today:

1. Globalized production, free trade, and offshore labour
2. The introduction and popularity of synthetic fibres
3. A shift in fashion and consumer preference to softer fibres and machine-washable yarns

I have read most of the available literature about the history of Canadian wool production, the majority of it focused on wool textile production—the cleaning, carding, spinning, weaving, and fulling of wool to make bolts of fabric which are then transformed into clothing and other textiles by the apparel and home-goods industries. Yarn for hand knitting was cleaned, carded, and spun for personal use, sometimes by hand in the home, and—as the population grew—in factories focused on producing yarn for knitters, crocheters, and weavers.

For accuracy, I have tried to differentiate between the wool textiles/apparel industry and wool manufactured into yarn for handicrafts, but because there is both crossover and a lack of differentiation in the literature, I haven't always been able to.

WHERE DID ALL THE MILLS GO? A SHORT HISTORY OF WOOL TEXTILE MILLS IN CANADA

The manufacture of cloth and clothing has played an important part in Canada's manufacturing, trade and architectural history. Textile mills provided employment for many during the late nineteenth and early twentieth centuries, and the production of clothing and cloth was ranked third in gross value of production in the early nineteen hundreds.[3] The legacy of textile mills can be seen in the development of international trade and tariff agreements and it even influenced architectural design in Ontario and Québec. Despite there being very little manufacturing of clothing and cloth in Canada today, its historical influence can still be seen.

> The story of textiles in Canada does not start with colonization; however, according to records sheep did not exist on this land base until the early 1600s, and I will only be covering the textile history that starts with wool production. We must consider the impact that agriculture and manufacturing have had on the land and the Indigenous people who were displaced and harmed by our clothing industry, past and present.

Early Days of Wool Production

In 1604, settlers first brought sheep to what we now call Canada. Sheep provided colonizers with meat, milk, and wool.[4] At this point in our textile history most woollen clothing would have been either brought over with the settlers or produced by hand in the home using small-scale processing equipment like hand cards, spinning wheels, knitting needles, et cetera. Knitting would have been a necessity, not a hobby,

---

3 A.B. McCullough, *The Primary Textile Industry in Canada: History and Heritage*, Minister of Supply and Services Canada, 1992.

4 François Castonguay, "Sheep Farming," *The Canadian Encyclopedia*, 28 May 2015, Historica Canada, www.thecanadianencyclopedia.ca/en/article/sheep-farming (Accessed March 2022).

and yarn would have been sourced from a family's own sheep or from a neighbour or community member.

By the end of the eighteenth century we start to see changes in the production of wool and textiles, with the introduction of large mechanically powered machines that shifted the production of wool from subsistence level to industrial scale and ultimately into the factory system we have today.

The first carding and fulling mills came to Upper and Lower Canada from Britain—the exact timing is challenging to pinpoint due to the lack of adequate record keeping at the time—but by 1803 there were three fulling mills in the Kingston/Napanee area, and mechanized wool and textile production had arrived.[5]

Growth of the Industry

The first census of Lower Canada was taken in 1827 and it provided the first accurate method to quantify growth in the woollen industry. At that time there were 91 carding mills and 79 fulling mills.[6] The majority of the spinning, weaving, and finishing was still carried out by individuals who were able to produce more fabric due to access to carding mills. Many would bring their wool to the mills to be carded, saving themselves considerable time over the hand-carding method. They would then take the carded fibres home to spin and weave. Then, in some cases, the fabric would be taken to a fulling mill to finish the fabric for sale.[7]

The impact and popularity of carding and fulling mills continued to grow into the 1840s and 1850s when new knitting machinery was introduced and the industry saw explosive growth, laying the foundation for the factory system of wool textile production.

By 1842 there were 186 carding mills and 144 fulling mills in Upper and Lower Canada.[8] The majority of weaving was still being done on

5 Ibid.
6 Ibid.
7 Government of Canada, *Report of the Royal Commission on the Textile Industry*, Justice W. F. A. Turgeon, 1938, Government of Canada Publications, https://epe.lac-bac.gc.ca/100/200/301/pco-bcp/commissions-ef/turgeon1938b-eng/turgeon1938b-part1-eng.pdf.
8 A.B. McCullough, *The Primary Textile Industry in Canada: History and Heritage*, Minister of Supply and Services Canada, 1992.

looms by home-based businesses. In 1851, hand-loom fabric production was 6.5 million yards, and this rose to 7.6 million yards by 1871.[9] I'm not much of a weaver, but that number is staggering to me.

In 1866, the Canadian-American Reciprocity Treaty—the first trade agreement between Canada and the United States (enacted in 1854)—was abandoned by the Americans. This significantly affected trade between the two countries but had a surprisingly positive impact on Canada's wool industry. Following the treaty's cancellation, the United States imposed steep duties on Canadian wool, leading to a surplus within Canada. This surplus fuelled rapid growth in the wool industry, as mills began using the excess to manufacture tweeds and flannel fabrics. As production expanded, new mills opened, creating more local jobs and further boosting the industry.[10]

By 1871 there were 270 woollen mills and 650 carding and fulling mills in operation. Collectively, they employed more than 4,443 workers and paid $917,287 in wages that year. The net value of production was more than $2 million (equivalent to around $52 million today).[11] Wool textile production was on the rise and was about to experience its biggest boost yet.

But by the late 1870s Canada was feeling the negative impact of the failed Reciprocity Treaty and the new tariffs on Canadian goods in the United States. At the same time, cheaper goods were entering the Canadian market from Britain, making it more challenging for Canadian businesses to compete. The lack of any protectionist policies for textiles meant that British woollen products were flooding the market. The British had access to different types of wool and new styles that were more desirable than the coarse, heavy cloth being produced in Canada. In 1879 John A. Macdonald and the Conservative Party of Canada

9 Government of Canada, *Report of the Royal Commission on the Textile Industry*, Justice W. F. A. Turgeon, 1938, Government of Canada Publications, https://epe.lac-bac.gc.ca/100/200/301/pco-bcp/commissions-ef/turgeon1938b-eng/turgeon1938b-part1-eng.pdf.

10 Ibid.

11 Government of Canada, *Census of Canada 1870-71*, Vol. 3, Department of Agriculture, 1873, Government of Canada Publications, https://publications.gc.ca/collections/collection_2016/statcan/CS98-1871-3.pdf.

introduced the National Policy, an economic and political strategy that placed higher tariffs on foreign imported goods with the goal of protecting Canadian manufacturers from outside competitors. Macdonald wanted to use the policy as a way to develop the country's economy, but some argue it was a political move to benefit wealthy businessmen who would continue to support the Conservative Party.[12]

For the Canadian wool manufacturing industry, the National Policy meant new mills were built, existing mills invested in larger machinery, and the tariff on any imported woollen goods was increased from 17.5 percent to 25 percent.[13] Growth in the industry was significant, and some historians call this the "heyday" of woollen manufacturing in Canada. By 1890 more than 7,000 Canadians were employed in over 377 woollen mills, and the net value of production was $4 million, which had doubled since 1870.[14]

Other factors also contributed to this period of growth. The Canadian population doubled in size between 1870 and 1914 and our textile consumption grew by almost 400 percent.[15] There was an influx of capital from large investors, and mills were able to grow and expand. And finally, the Canadian Department of Militia and Defence contracted significant amounts of wool fabric from the various mills to produce blankets and clothing for soldiers, including those fighting the Boers in South Africa in 1899 and those fighting against the Métis Resistance in 1885.[16]

12 Edward Porritt. "Canada's National Policy." *Political Science Quarterly*, June 1917, Vol. 32, No. 2 (June 1917), pp. 177–208, https://www.jstor.org/stable/2141729.

13 Government of Canada, *Report of the Royal Commission on the Textile Industry*, Justice W. F. A. Turgeon, 1938, Government of Canada Publications, https://epe.lac-bac.gc.ca/100/200/301/pco-bcp/commissions-ef/turgeon1938b-eng/turgeon1938b-part1-eng.pdf.

14 Government of Canada, *Census of Canada 1870-71*, Vol. 3, Department of Agriculture, 1873, Government of Canada Publications, https://publications.gc.ca/collections/collection_2016/statcan/CS98-1871-3.pdf.

15 A.B. McCullough, *The Primary Textile Industry in Canada: History and Heritage*, Minister of Supply and Services Canada, 1992.

16 "A Chronology of the Textile Manufacturing Industries of Canada, 1883 - 1983," *Canadian Textile Journal*, June 1983.

Decline of the Industry

This growth in the woollen textile industry did not last. Many factors that contributed to its decline, including reduction in tariffs on British fabrics, changing styles, the availability of finer wools, and growing urbanization in Canada. In 1897 the British Preferential Tariff came into place, effectively reducing the tariffs on British woollen goods by a third. This, combined with changing fashions and customers wanting finer wool fabrics, meant that Canadian mills had to import more wool to meet the demand. The heavy coarse woollen goods of the past were no longer desirable by the burgeoning urban population in Canada. Local, rural mills couldn't compete with the large British manufacturers and they didn't have the capital to pivot their operations. By 1910 there were only 87 textile mills remaining and the industry was struggling to survive.[17]

World War I brought a brief upswing in wool production, and increased mill capacity was needed to supply the military with wool textiles. This was not to last, though, as the post-war boom meant more foreign textiles coming in and the introduction of the first synthetic materials. In the 1920s two rayon plants were built in Cornwall and Drummondville, Ontario, and by the 1930s the demand for rayon had quadrupled.[18] The production of rayon was seen as a "triumph of synthetic chemistry combined with modern engineering skill" to produce a synthetic material that could rival natural fibres.[19]

The textile industry in Canada continued to have ups and downs in terms of production, employment, and its share in the economic development of Canada. However, wool textile production was replaced

17 Government of Canada, *Report of the Royal Commission on the Textile Industry*, Justice W. F. A. Turgeon, 1938, Government of Canada Publications, https://epe.lac-bac.gc.ca/100/200/301/pco-bcp/commissions-ef/turgeon1938b-eng/turgeon1938b-part1-eng.pdf.

18 A.B. McCullough, *The Primary Textile Industry in Canada: History and Heritage*, Minister of Supply and Services Canada, 1992.

19 Government of Canada, *Report of the Royal Commission on the Textile Industry*, Justice W. F. A. Turgeon, 1938, Government of Canada Publications, https://epe.lac-bac.gc.ca/100/200/301/pco-bcp/commissions-ef/turgeon1938b-eng/turgeon1938b-part1-eng.pdf.

by increased synthetic fibre production and cotton production using imported cotton.[20] The woollen mills were replaced by knitting mills and then wool was replaced with other fibres.

The Second World War brought about brief growth due to controlled wages and subsidies for manufacturing. However, after the war, labour unions and workers advocated for increased wages and better working conditions. At the same time, international trade was beginning to gain popularity through the development of the General Agreement on Tariffs and Trade (GATT), which allowed for greater amounts of imported wool goods. With the push for higher wages and the increase of imported wool goods, wool mills struggled to compete.[21] And since wool production is more labour-intensive than synthetic-textile production, trends continued to favour synthetic production. The wool manufacturing industry continued to lose its position in Canadian textile manufacturing.

The 1960s led to new challenges that have come to define not only the modern wool production industry in Canada, but our entire clothing and textile industry. During this time, Hong Kong, Taiwan, and South Korea emerged as manufacturing hubs due to their large pools of low-cost labour and less rigorous labour protections. Canada negotiated, and then implemented, quotas with these countries to limit their imports. This eventually became the Multi-Fibre Arrangement.[22] A few years later, in 1989, Canada and the United States signed the Free Trade Agreement (FTA), a precursor to the North American Free Trade Agreement (NAFTA), which allowed for more free-flowing trade of clothing and textiles between the two nations. Then, between 1995 and 2002, the World Trade Organization (WTO) lifted the pre-existing quotas on clothing imports to all member countries. This meant a

20 Canada does not have the climate for growing cotton; it is imported as a raw material. It was spun in the first Canadian cotton mills in 1844 in Lower Canada, and production capacity continued to grow along with its popularity as a fabric choice for the Canadian population (McCullough, 1992).

21 A.B. McCullough, *The Primary Textile Industry in Canada: History and Heritage*, Minister of Supply and Services Canada, 1992.

22 Diana Wyman, "Trade Liberalization and the Canadian Clothing Market," *Canadian Economic Observer*, December 2006.

major influx of clothing (of all fibres) into Canada, and an increase in imports from those countries that could manufacture cloth and clothing at a far lower cost than Canada could. By 2005 China was a full member of the WTO and all remaining quotas on textiles and clothing had been removed; Chinese imports now accounted for 46 percent of all the clothing imports into Canada.[23]

I was twenty-one years old in 1999 when the WTO Ministerial meeting happened in Seattle, Washington, and huge numbers of protesters showed up to protest the environmental, social and labour implications of the new WTO agreement. The protesters were met with overwhelming police brutality and violence, adding a North American fervour to the global anti-globalization movement.[24] I was not at the protests in Seattle, but I was at most large-scale anti-globalization protests that followed, until 2003. I was passionate about the urgency to stop the churning wheels of globalization and I was worried about what it would mean for the environment, labour, human rights, and local industries. Globalization has facilitated a worldwide textile industry that can take advantage of lower production costs, advanced technologies, increased efficiencies, and an enormous amount of output at an extremely low cost to the consumer. But the complexity of this system means that consumers are disconnected from the process of textile production and never stop to ask, "At what cost did I actually buy this shirt?" They don't consider the true cost of polluting waterways or the impact of low-wage labour on community and family. They don't think about the cost when people die in a factory fire, or the cost of farmers not having access to education and health care.

Fast fashion (how I most often refer to our current clothing system) is a system characterized by the mass production of inexpensive and low-quality clothing that replicates the designs of popular fashion brands, and the frequent consumption of this clothing with no regard

23 Ibid.

24 Photo gallery of the protests in Seattle: https://www.seattle.gov/cityarchives/exhibits-and-education/digital-document-libraries/world-trade-organization-protests-in-seattle.

to the impacts of its disposal. This system is not inevitable. We have the power—with our dollars and our votes—to prioritize a clothing and textile system that accounts for the true costs of our clothing. For me, this future must include wool production and manufacturing.

This very brief look at the history of wool textile and clothing manufacturing in Canada provides one broad perspective on an issue that affects every Canadian, since we all must wear clothes. But now I want to shift gears and look at the history of the Canadian wool with a specific focus on wool yarn for fibre arts and the handcraft industry.

History of Yarn Mills in Canada

Knitting first came to Canada with settlers. Both France and England had knitting cultures, and as settlers arrived they brought needles, yarn, and eventually sheep. Early knitting yarns would have been from French or English sheep breeds, until sheep were introduced to the land base and settlers had access to "homegrown" wool. We can look through our history and see that knitting contributed to our material culture as the country was settled and developed. Whether it was the iconic trigger-finger mittens from Newfoundland, Arctic socks worn by fishers in Nova Scotia, the circular shawls—called "tippets"—worn by women in the 1840s or the "tam-o'-shanter" hats worn by kids and hockey players alike—knitted clothing has been both a practical and emblematic part of our culture for hundreds of years.[25]

Early access to wool was community-based and the industry grew as economic opportunities presented themselves. The story of hand knitting in Canada can best be explored through the stories of the classic brands and mills that many see as synonymous with yarn. Like the wool textile mills, the knitting yarn industry has also been impacted by globalized supply chains, the introduction and popularity of synthetic fibres, and cheaper offshore labour. It has also seen legacy mills and family-owned companies weather these storms to maintain their Canadian milling history.

25 Shirley A. Scott, *Canada Knits: Craft and Comfort in a Northern Land*, McGraw-Hill Ryerson Ltd., 1990.

Briggs & Little, started by George Lister on the Magaguadavic River in New Brunswick in 1857, was originally known as York Woolen Mills. It was bought in the 1890s by Roy Little, who operated the mill until it was purchased in 1916 by Mathew Briggs and Howard Little (no relation).[26] The mill has been owned by the descendants of those owners ever since and has always purchased and processed wool grown by Canadian sheep farmers. The mill has been rebuilt four times due to devastating fires, and was partially powered by water until 1994. Briggs & Little started with only a handful of coloured yarns: black, white, and four shades of grey. Over time they have grown their colour lines, introduced numerous different types of yarn, and have maintained their positioning as a truly Canadian wool mill, spinning and selling 100 percent Canadian wool. Currently, they purchase wool direct from Canadian sheep producers, and graded wool from CCWG.[27]

The operation is a testament to the scale of production that is necessary to adequately transform large quantities of Canadian wool into products. Briggs & Little has never used offshore labour or manufacturing and has prioritized local production, local employment, and demonstrated the economic viability of a larger-scale mill in Canada. According to current owner Leah Little, their "quiet success" for the past 167 years is due to a loyal customer base, knowledgeable staff, and a commitment to their local community. All their processing is done on site and they do not outsource any aspect of the business.

Western Canada—Mary Maxim, White Buffalo Yarn, and Cowichan Sweaters. The history of Canadian wool production was centred in Ontario, Québec, and the Maritime provinces, but Manitoba has a unique history of wool mills and craft yarn production that is worth highlighting, as is its connection to the Pacific coast.

We can't talk about Canadian wool yarn without talking about Mary Maxim, an iconic yarn and pattern company that started in the 1930s in Sifton, Manitoba. Olive and Willard McPhedrain started

26 Briggs & Little Website, https://briggsandlittle.com/history/ (Accessed October 2024).
27 Interview and email correspondence with Leah Little, November 2024.

a business called Sifton Wool Products Ltd. The main products were blankets, wool batting, knitted work socks, spinning wheels, and other tools. The name was changed over time to Mary Maxim as a marketing tool to promote the homemade work socks that were knitted by an employee, "Miss Mary Maximchuk."[28] While sales of their products were adequate, it wasn't until the company started producing their "graph style patterns" that things really took off for Mary Maxim. I'm sure many of us who grew up on the Prairies owned at least one sweater made from a Mary Maxim pattern—and if you are a child of the '70s or '80s, chances are that sweater had a Sesame Street character on the back. (True story—we still have the Mary Maxim Cookie Monster sweater that my husband's grandma knitted for him).

The inspiration for this style was actually the Cowichan sweaters created by women from the Cowichan Nation on Vancouver Island, British Columbia. The development of these patterns launched the company and its yarn into a new era of manufacturing, and soon they partnered with Ontario-based mill, Spinrite Yarns, to produce the Northland and Narvik yarns that became staples for many Canadian knitters. In 1954 Mary Maxim moved from Manitoba to Paris, Ontario, where it still exists today.[29] It is hard to say for certain how much of their yarn was made with Canadian wool versus imported wool. One advertisement in the *Manitoba Cooperator* from October 1953, promoting the Northland Sweater wool, states, "Heavy homespun is made with pure 100 percent native wool, with long-fibred New Zealand wool added for extra strength."[30] Mary Maxim was manufacturing the wool yarn in Canada, but was already using imported wool for their products. Mary Maxim refused to comment on how much of their wool is imported, either historically or currently, or what percentage (if any) of their processing is done in Canada; their head office is now in the state of Michigan. They continue to sell their own brand of yarn and their

28 Shirley A. Scott, *Canada Knits: Craft and Comfort in a Northern Land*, McGraw-Hill Ryerson Ltd., 1990.

29 Ibid.

30 *Manitoba Co-operator*, October 22, 1953, https://www.manitobacooperator.ca/country-crossroads/our-history/our-history-october-1953/.

iconic knitting patterns, but I have no way of knowing if any of it is sourced or processed in Canada.

Finally, there is the complicated and fraught history of the White Buffalo mill, that popularized knitting among Western Canadians from the 1950s until the early '90s. This mill in Brandon, Manitoba, started as a lumber mill and then became the Brandon Woolen Mill. By the 1960s it was called the Metev Woolen Mills, and then it was eventually renamed White Buffalo. The mill produced woollen spun yarn and in the beginning they produced blankets from scrap woollens that were brought in for re-processing. According to one report, the mill produced 300,000 lb. of wool yarn a year, plus blankets![31]

In the early 1960s the owner, Chris Metev, came across the original Cowichan sweater and was captivated by the style and construction, but most intrigued by the chunky unspun wool that the Indigenous makers had carded by hand and then used in the knitting of the sweaters. He decided that his machinery could produce a similar yarn and save significant labour time. This is how the "White Buffalo unspun yarn" was created. It is now often produced as a six-strand unspun yarn packaged as a large round "cake," but it was originally one or four strands. Mr. Metev patented the process immediately and the market demand for this wool dramatically grew in the Prairies.

In the 1980s, the White Buffalo design and pattern department sponsored a national contest and encouraged the use of their unspun yarn in the designs. They published almost 100 sweater designs that clearly copied the traditional Cowichan style.[32] They continued to produce their White Buffalo yarn using exclusively Canadian-grown wool until the late 1990s, when the mill closed down. There is no verifiable information on what happened to this mill or its equipment when it closed.

White Buffalo, Mary Maxim, and authentic Cowichan sweaters are often considered synonymous by wearers and even by knitters. But according to knitwear designer and Cowichan sweater historian Sylvia

31 Ibid.
32 Ibid.

Olsen, they are all very different, even if they inspired each other. The long-term effect of the White Buffalo patterns has deeply hurt the economic prosperity of the makers of authentic Cowichan sweaters, and the economic development of the Cowichan Nation.[33]

According to Sylvia, "Companies hired hand knitters, mostly women who were recent immigrants and willing to work for extremely low wages, to produce sweaters commercially from White Buffalo patterns. The White Buffalo products suppliers provided a one-stop marketing opportunity for Cowichan sweater shops, offering quantities of sweaters with standardized sizing, quality, and designs, and eliminating the time-consuming process of buying one sweater at a time from individual Coast Salish knitters. Pretty soon, retail stores took copying Cowichan sweaters one step further. They sold Cowichan look-alikes next to the genuine sweaters, often calling the copies 'Cowichans' and undercutting the price and confusing consumers." (Olsen, 2021)

Many have worked to tell the Cowichan sweater story and to define the authentic style of this sweater and the economic impact it had on Indigenous communities. Although Mary Maxim is still around and still selling their sweater designs that took inspiration from the original Cowichan sweater, there has been little accountability or reconciliation for the ways in which the Cowichan sweater was copied, altered, and turned into profit for non-Indigenous companies. Cultural appropriation is an ongoing topic of discussion for pattern designers, yarn companies, and small businesses as we navigate the intersection of historical and cultural design, and contemporary inspirations.

Patons & Baldwins. If you walk into a Michaels craft store today you can most likely find Patons wool and patterns with the legendary beehive logo. Before Ravelry.com and "stitch 'n bitch" and Debbie Bliss, we had Patons patterns. The company Patons & Baldwins is originally from England, where it started as a textile mill in the 1780s and expanded in the twentieth century beyond its English borders. Originally a mail-

33 Sylvia Olsen, *Unravelling Canada: A Knitting Odyssey*, Douglas & McIntyre, 2021.

order business through the Eaton's catalogue in Canada, they eventually began looking for a local manufacturing operation. They found one in 1928 when they acquired a mill from Thomas Aked.[34]

Thomas Aked immigrated to Canada in 1908 and became the manager for the Monarch Knitting Company. In 1918, he used that experience to formed Aked & Company Ltd. in Toronto—a factory that produced "fancy hand-knitting yarns."[35] Aked owned and operated the factory for a decade until Patons purchased it and built a larger mill in 1931. It wasn't until 1963 that Patons officially incorporated in Canada. They continued to manufacture and produce yarn, but they imported more and more wool from elsewhere, and began using more synthetic fibres (acrylic and nylon).

Patons patterns and yarns are still popular on the shelves of the more commercial craft stores, and they continue to be a significant part of the yarn industry today, with a variety of wool and synthetic yarns. They are now distributed in Canada by Spinrite Yarns, a company with its own big history of manufacturing in Canada.

Spinrite Yarns had its start in Canada in 1913 when Max Becker, a German immigrant and qualified machine knitter, met Aaron Ringler. Ringler convinced Becker to check out the town of Listowel, Ontario, because of its access to fresh water and rail lines. This serendipitous meeting resulted in the establishment of the "Perfect Knitting Mills" that manufactured machine-knit sweaters, many of which were supplied to the Canadian Armed Forces for decades.[36] The mill was sold in 1931 and renamed Maitland Spinning Mills Ltd., after which it struggled to maintain its market share, but in 1952 it was purchased by David Hay, who renamed the company Spinrite. Shortly thereafter, while Mary Maxim sweaters were growing in popularity, Spinrite developed a heavy homespun-type yarn they branded "Northland

34 Shirley A. Scott, *Canada Knits: Craft and Comfort in a Northern Land*, McGraw-Hill Ryerson Ltd., 1990.
35 Michael Harvey, *Patons: A Story of Handknitting*, Springwood Books Limited, 1985.
36 Ibid.

Yarn," which used some Canadian wool and some imported New Zealand wool.[37] Spinrite worked hard to manufacture a variety of products that met consumer demands for the time. They had an in-house dyeing operation and modern yarn spinning equipment that could do worsted and woollen spinning. They manufactured both wool and synthetic yarns for both the hand-knitting customer and the machine knitting industry.

According to their website, Spinrite mill has become the largest distributor of yarn, thread, and needle craft accessories in North America. They have bought out some of the largest brand names in commercial knitting, including Red Heart, Bernat, Caron, Patons, Coats & Clark, Phentex, and more, which they distribute to the large commercial craft stores throughout Canada. Although they still have a large manufacturing operation in Listowel where they manufacture their yarns, they do not work with any Canadian wool and instead import all the wool they spin.

TRANSPARENCY AND TRACEABILITY

Although Spinrite still maintains its Canadian roots, the traceability of the wool it uses is not readily available. This scenario is similar to some of the other large yarn companies that you may recognize on the shelves of smaller, independent local yarn stores—brands like Berroco, Cascade, Lang Yarns, Manos del Uruguay, BC Garn, Debbie Bliss, Sirdar, and others. Many of these companies are working to limit the environmental cost of production and to provide more information about the value chain of their products. Some of their yarns are even certified under the Global Organic Textile Standard (GOTS) or OEKO-TEX® Standard. But how do we, as consumers, access information about these companies and how do we find out what, if any, certifications they have and what that information—or lack thereof—means in terms of the impact or true cost of these products?

37 Manitoba Co-operator, October 22, 1953, https://www.manitobacooperator.ca/country-crossroads/our-history/our-history-october-1953/.

During my research for this book, I reached out to many of the yarn companies mentioned in this chapter. I asked questions about what farms, or even what countries, they purchase their wool from. I asked what mills they use, and if they knew the labour or environmental standards required of those mills, and if they had any information about the waste created from the process of transforming wool into their yarns. I asked if they had traceability information in place that they could easily share with consumers.

Many companies simply didn't respond, despite numerous emails and phone calls. Those who did respond often answered with a blanket statement about the global certifications that their company subscribed to, or at least that some of their products were included within. For proprietary reasons they would not, or could not, share any more information about the mills they use, but most assured me that the labour and environmental standards of the country the mills were operating in were followed. That doesn't necessarily mean that those laws are what is best for people or the planet, but we can't know for sure because we don't have enough information. Further to this, many of the companies requested that their names be excluded from this book.

The exceptions to these leery responses came from Canadian companies like Briggs & Little, Custom Woollen Mills, and the many small mills I contacted, who were more than happy to share the details of the wool they use, their processing, and their desire to stay rooted in community. The dichotomy between the large companies that use protectionist language to avoid transparency and the companies that are hiring Canadian workers and building relationships with Canadian consumers was very telling. Do we want to support companies that are open to sharing their business practices and the impact they have on community and environment, or do we want to support companies that are not willing to be transparent about the impact they have on the environment and workers?

I don't believe that there is some corrupt large-scale wool plan out there, or that this lack of transparency is hiding something nefarious. But I do think the lack of transparency reveals the priorities of these companies and that as consumers we must push for more answers. If we don't ask for the traceability information for our yarn—how and where it is made and who is impacted by its production—then why would they be compelled to share? As consumers we need to start asking yarn companies, local yarn stores, and even big chain craft stores about the traceability information of the yarn they carry. We need to push for more transparent labelling on our wool products so we can form a clearer picture of their true costs. And finally, we need to support those companies that openly share the true costs of their production—many of these companies are those that are established in our local cities and towns.

PROCESSING WOOL IN CANADA TODAY

One morning in 2012, I was working in my community-based yarn shop, Baaad Anna's, in Vancouver, when one of my "regulars" came into the shop to tell me about the amazing wool tour she went on in Prince Edward Island. I'm sure I nodded and smiled and feigned interest while trying to just keep my eyelids open—I had just had my second baby and was barely keeping it all straight. Then she handed me a large, laminated poster and said something like, "I think this is what you've been looking for; it was this incredible place that I visited in Belfast, P.E.I., and it made me think of you and everything you talk about with local wool." I genuinely thanked her for thinking of me and for the thoughtful gesture and said I would be sure to read through the poster—sadly I did not. It ended up in the piles of stuff that I just didn't have the bandwidth to deal with at that time.

Three years later I sold the store and we packed up our kids and travelled to the centre of Canada to start a sheep farm. It was a move so full of hubris that I'm slightly embarrassed to think back to that time—thankfully the many missteps I made that year firmly put me back in my place! In January of 2017, I opened up the last untouched box from our big move—I was looking for a knitting pattern that I knew for certain I had brought with me and I was finally ready to start knitting it (I know it's weird in the age of Ravelry and so many other online knitting apps to actually have a paper knitting pattern, but I really prefer to follow a printed pattern). I started digging through the box that was mostly paperwork from my time at Baaad Anna's and as I got closer to the bottom of the box I pulled out that laminated poster from 2012. It was the most incredible thing I had ever seen. It was from Belfast Mini Mills. The poster explained all the machines and machinery that they manufactured and sold as a "micro-mill"—an entire wool processing mill that could fit into a 24-by-24-foot space and transform wool into yarn. I raced up the stairs and dropped it in Luke's lap and said, "THIS—this is exactly what we've been looking for—we need to call them right now!" I put a pink sticky note on the poster that said, "This is the plan." And by mid-summer 2018, we had a Belfast Mini Mill set up in our garage on the farm.

Regardless of the size, or equipment, or age of the wool processing mill, all wool mills perform the same functions. Some mills can be run by one to three people and some require ten to twenty people to function. But if you toured any of these mills—and I definitely encourage you to tour a wool mill if you ever get a chance—you would find machinery to accomplish these tasks, (in order):

Grading/Assessing/Skirting/Sorting. This is the first step in wool production, where the wool is evaluated based on its cleanliness (how much vegetable matter contamination there is), staple length, crimp, colour, and micron count. This is an important first step to ensure the highest quality of wool ends up being processed (lower quality wool is often discarded).

Scouring. Next, the wool is washed to remove any dirt, contaminants, and most importantly, lanolin. (Lanolin is the greasy substance excreted from the pores of sheep that protects the wool and the sheep's skin; it can be used in the beauty/cosmetics industry and must be removed before mechanically processing wool). Scouring requires hot water and a detergent that binds with the grease to remove it from the fibre. The final part of this step is the post-scouring, which involves the drying of the fibre and the removal of the waste effluent.

> *Scouring is the essential link between agriculture and industry. It is the necessary step that transforms wool from an agriculture by-product to a value-added commercial or industrial product. Scouring is also one of the largest bottlenecks challenging smaller mills across the country.*

Small and medium mills can often process more wool than they can wash. There is only one large scouring operation in Canada; many mills would benefit from regional scouring facilities.

Picking & Carding. This stage is sometimes done before scouring, and sometimes after. Picking opens up the fibres and separates the wool locks to prepare them for carding. Carding is the process of separating the individual fibres and combing them to form a continuous web for batting or roving.

Spinning. There are many different techniques and mechanical processes to take carded wool to spun yarn, but essentially the fibre is drawn out and then twisted into a single strand of yarn and wound onto a bobbin. From there they can be plied with one or more other single strands of yarn by twisting them in the opposite direction. The finished yarn is then wound onto cones or into skeins, balls, or cakes to be used by hand crafters or in mechanical textile production.

Regardless of the size of the mill, all wool will go through these steps to be made into yarn.

How Many Mills Do We Have in Canada?

"I don't know why you would want to start a wool mill. It is hard, dirty work and the profit margin isn't there. No one gets rich running a wool mill." This was the first response in my email inbox after I reached out to some wool mill owners in 2017, when I was considering opening up our mill. I wanted their advice and recommendations—and their encouragement. But most of the responses were like this—incredulity at the very suggestion of starting this kind of business.

I didn't let it sway me and years later I'm very thankful that I forged ahead—but I can also see their point. My husband always says, "If you're looking to get into a business just to be self-employed, you'd be better off starting a food truck. There are faster, easier ways to make money than starting a wool manufacturing business in Canada." He isn't wrong, but our goals for our farm and mill are more than just profit generation. There are quite a few more mini- and small-sized wool processing mills across Canada today than there were when we started ours in 2018—demand is growing and the willingness of folks to take the risk to meet that demand is spreading.

For fibre farmers who want to get their wool processed, they have some options in most provinces in Canada (at the time of writing this book, there are no fibre processing mills in the northern territories). There are three larger mills that are focused on processing only Canadian wool: Custom Woolen Mills in Alberta, Briggs & Little in New Brunswick, and MacAusland's Woollen Mills in Prince Edward Island. These mills have incredible history and have forged their own paths through the Canadian wool industry by prioritizing local production, rural employment, and strong customer loyalty. In general, each of these mills can process between 75,000 and 150,000 lb. of wool in a year.[38]

The other mills across the country are a mix of mini mills (most often the small-batch processing equipment from Belfast Mini Mills), or mills employing mix-and-match equipment to process batting, roving, or yarn. There is at least one small mill in every province across the country, with the highest concentration of wool-processing equipment in Ontario. Most of these mills offer custom processing services for fibre farmers. Generally, each of these mills can process between 1,000 and 10,000 pounds of wool a year.[39]

---

38 Approximate amounts based on correspondence with large mills in November 2024.

39 Amounts self declared by Canadian Mills in the Canadian Wool Collective Mill survey.

There are also still a few mills in Canada that do larger-scale spinning, dyeing, felting, and finishing—these mills are part of a larger supply chain where much of the raw materials are imported, processed on site, and then transformed into finished goods to be sold. Filature Lemieux Spinning is a family-owned and -operated large-scale mill in Québec, founded in 1906 by Joseph-Albert Lemieux to card wool for his local community. The mill grew steadily for sixty years, adding new equipment and processing to adapt to changing demands. They began processing exclusively Canadian wool and expanded to wool from New Zealand to produce a good quality carpet wool. The wool they imported had less vegetable matter and was more consistent in larger quantities than what they had access to in Canada.

In the 1980s, circumstances changed for Lemieux and the mill pivoted in some key areas. The first was that they shut down their in-house scouring operation, which was going to require a new effluence waste treatment facility. This necessitated a shift to only being able to process pre-scoured wool, most of which they sourced from New Zealand. They also saw an increase in consumer demand for synthetics, and they became an industry leader in synthetic and synthetic-blend yarns. The mill operated with success for thirty years with their supply of New Zealand imported wool and synthetics until 2019, when they noticed the industry was starting to change again toward local Canadian wool.[40]

With increased import pressure, and consumer demand shifting to local wool, they began looking at niche markets and once again sourcing Canadian wool. Because they can only work with pre-scoured wool, many of their clients utilize scouring facilities in Ontario and use Lemieux for carding, spinning, and dyeing. Lemieux is keen to work with others that are focused on a local supply chain and want to improve the traceability of their products. The mill currently spins about 15,000 lb. of wool per week, but they have the capacity to spin more. Lemieux will spin custom batches of local Canadian wool, but the minimum batch size is 1,000 lb. and the wool must be pre-scoured.

40 Interview with Serge Lemieux, February 10, 2025.

They continue to work with synthetic fibres and import some wool from New Zealand and England, but they see the future in Canadian-grown and -manufactured wool. During my interview with Serge Lemieux, the fifth generation to own and operate the mill, he expressed his hope for a new Canadian wool industry: "There is a way to get a better value for the wool if good care is taken on it. Finding the proper value of the product depends on where it goes. And this is our priority now." Elitespun/Amtex operate a large-scale mill in Ontario. In addition to Lemieux, theirs is one of the last remaining large-scale spinning operations still running in Canada. They do not work with Canadian wool and declined to comment on the past, present, or future of working with Canadian wool.

VISION FOR A UNIFIED WOOL INDUSTRY

There is an unfortunate disconnect between the artisanal fibre arts industry and the larger commercial/industrial wool industry. Although the size of the artisanal fibre arts industry is perceived to be insignificant compared to the industrial wool industry, it is still a multimillion-dollar industry in Canada. In my experience, small-scale and industrial-scale producers seem to be pitted against each other and neither fully understands the other in terms of output or future market potential. I believe that attempts have recently been made to connect and unify the efforts of these industries to grow their markets, and to recognize the value in both small-scale artisanal work and larger-scale applications of domestic wool. My hope is that a more unified wool voice in Canada would include commercial farmers, small-scale mill owners, industrial mills, and small-scale farmers as well. It is important to consider that when the industry set its sights offshore and on synthetics, it was the small-scale farmers and a handful of small/medium artisanal mills that kept the Canadian wool industry alive. At the same time, the artisanal sector has much to learn from the strategic thinking of industry-led organizations.

CHALLENGES IN THE CANADIAN WOOL PROCESSING INDUSTRY

"Can you imagine telling egg farmers or dairy producers they have to wait twelve months before they can get their agricultural product to market? Wait times are killing the small-scale wool farming industry in Canada." I wrote these words on a grant application for more mill equipment. Perhaps it was a bit dramatic, but I wanted it to be shocking, and hopefully to encourage the provincial government to approve the grant.

Despite growth in the small-scale wool processing industry, we are still not able to meet the demand of fibre farmers in a timely and efficient manner. The existence of small/mini mills has in many ways revolutionized the small-scale fibre farming industry in Canada. The flexibility and capacity to manage small amounts of fibre and perform small-batch processing makes possible a higher-end product that appeals to those looking for unique yarns, breed-specific wool, and completely traceable wool products. It is the missing piece of infrastructure in the wool value chain for fibre farmers.

That being said, fibre farmers often have to wait twelve to eighteen months to get their wool processed. Farmers have to ship or drive their wool to a mill that is a significant distance away, and the cost of small-scale processing can be prohibitive due to its lower output compared to large-scale processing and specialized labour requirements.

I have spent considerable time talking with other mill owners and advocates for the wool industry about growth in the manufacturing sector. How can we grow a regional wool-processing industry that performs the necessary functions at a reasonable price for customers while still paying living wages and attracting knowledgeable workers? How do we educate consumers about the difference between local manufacturing and offshore processing, and the price disparity? And how do we scale up domestic processing to move beyond artisanal or small-batch processing to adequately meet the needs of larger Canadian sheep farms?

THE TRUE COST OF DOMESTIC WOOL PRODUCTION

Processes and prices vary slightly from mill to mill, as do the services offered. Custom processing service is when a farmer or artisan brings their wool to the mill and it is transformed into a value-added product that they will sell or use. Alternatively, mills buy raw wool from farmers and transform it into value-added products that they sell directly to customers. Some mills do not offer custom processing at all anymore because the numbers just don't make it a viable business opportunity; the return on investment in creating retail yarn is much higher. Each mill has its own specific policies and pricing for processing wool.

Mills will often have a minimum weight requirement for processing, which can be restrictive or beneficial depending on the producer. The larger mills often have larger minimum processing amounts to justify the time and energy requirements of their machinery, whereas the mini mills have the capacity to do smaller batches of a couple of pounds, or fleece-specific yarn products while still being able to cover costs. On the other hand, the mini mills have a more challenging time doing larger batches of hundreds of pounds, and the larger mills can offer a lower price per pound for large batches.

Some mills charge a per-kilogram or -pound price at every stage of processing, and some charge an aggregate price. Wool generally loses anywhere from 30–60 percent of its weight during the various stages of processing—most of it at the scouring stage—so some mills will break down the price based on that changing weight.

FOLLOW THE WOOL

Here is the breakdown of processing our three scenarios of Canadian wool into a worsted weight 3-ply yarn at a local mill (see page 34 for the three scenarios).

Scenario 1: 5 lb. coloured Shetland wool

- Scouring: $12.00/lb. = $60.00
- Carding 3 lb. of wool (40% weight loss at scouring): $18.00/lb. = $54.00
- Processing 2.5 lb. of wool (10% loss in carding) into 3-ply worsted weight yarn: $21.00/lb. = $52.50

Total cost of processing = $166.50
Yield: 11.3 skeins
Cost per skein: $14.73

Scenario 2: 8.5 lb. fine wool

- Scouring: $12.00/lb. = $102.00
- Carding 5.1 lb. of wool (40% weight loss at scouring): $18.00/lb. = $91.80
- Processing 4.25 lb. of wool (10% loss in carding) into 3-ply worsted weight yarn: $21.00/lb. = $89.25

Total cost of processing = $283.05
Yield: 19.2 skeins
Cost per skein: $14.74

Scenario 3: 8 lb. medium wool

- Scouring: $12.00/lb. = $96.00
- Carding 4.8 lb. of wool (40% weight loss at scouring): $18.00/lb. = $86.40
- Processing 4 lb. of wool (10% loss in carding) into 3-ply worsted weight yarn: $21.00/lb. = $84.00

Total cost of processing = $266.40
Yield: 18.6 skeins
Cost per skein: $14.32

Summary of Costs to Process 100 g of Wool in a Small Canadian Mill

***Scenario 1:* $14.73/100 g skein**
***Scenario 2:* $14.74/100 g skein**
***Scenario 3:* $14.32/100 g skein**

The cost of production for each individual skein will vary depending on the mill chosen, the breed of sheep, and the weight classification of yarn. For example, it takes much longer to produce a fingering weight yarn than a worsted weight yarn, so the cost of production has to reflect that time investment. There are other factors that affect the cost of small-batch processing; for example, scouring fine wool often takes an extra wash that will increase the cost of production.

The production cost of $14/skein results in locally manufactured wool having a higher base price point than the commercially available brands that are manufactured in a country with larger equipment and cheaper labour. And this cost doesn't yet include the cost of growing the wool and raising the sheep, or dyeing the wool, or the work involved to get it to market. Often, farmers and small producers subsidize their own yarn to make it competitive in the marketplace, or they price it as a higher-end yarn to justify the price to consumers who don't understand the true cost of local production. These local farm yarns are competing in a market where the base costs of the majority of commercial yarns are significantly lower.

UNDERSTANDING "MADE IN CANADA"

When I learned to knit, options for 100 percent wool yarn were limited, and the knitting industry of today was just starting to gain momentum.

When I opened up Baaad Anna's Yarn Store in East Vancouver I ended up stocking my shelves with the big names in commercial yarn at the time— Cascade, Debbie Bliss, Sirdar, Noro, Berroco, and more. I had one section for indie-dyed yarn, and I was thankful to have access to excellent local small-batch yarn from SweetGeorgia Yarns and Indigo Moon. By the time I sold my store and started farming just five years later, the industry had exploded with many more Canadian indie-dyed yarn businesses, and it continues to grow.[41]

I adore the creativity and the vibrancy that this growth has added to the industry. The beauty and expression of colour that is on display at every fibre festival and yarn show is a testament to the artistry and entrepreneurial spirit that is shaping our industry. Yet I am frequently asked by dyers where they can source Canadian-made yarn bases, and I have little to offer them as an answer.

Many of the indie-dyed yarns available today are made using the same base yarn—a blend of superwash Merino, sometimes blended with nylon, cashmere, or silk. Much of this yarn is sourced from a Canadian company, but the wool is not grown or sourced in Canada. I have seen different labels on this yarn—"Hand-dyed in Canada," "Made in Canada," "Dyed in Canada," "Small-batch Canadian dyed"—all focused on differentiating themselves from one another, but ultimately featuring the exact same base yarn made from wool and other fibres that are not grown in Canada.

In Canada there is one main supplier for most of the base yarn used by indie/hand-dyers. I reached out to this supplier to find out about the traceability of their wool, and they provided some information, under the condition that I do not name the company. They are a spinning mill that operates in Canada and therefore is subject to the labour and environmental laws that govern Canadian manufacturing. They exclusively import wool for their products and do not work

41 Eva Koronios, "Fabric, Craft & Sewing Supplies Stores in Canada Doing Sew Well: Increased Operating Efficiency Among Leading Operators Is Anticipated to Preserve Industry Profit," *Craft Industry Alliance*, May 2020, https://craftindustryalliance.org/wp-content/uploads/2021/11/May-2020-Fabric-Craft-Sewing-Supplies-Stores-in-Canada-Industry-Report.pdf.

with Canadian-grown wool. This wool has been scoured (washed) and carded prior to being imported, and they do the spinning and finishing stages in-house. They were not willing to share extensive information about the source of their wool and their supply chain due to the "proprietary information" that provides them "a competitive advantage as one the of the last spinning mills in Canada."[42] They did confirm that some of the wool they source is RWS certified and some is not. The Responsible Wool Standard (RWS) is an international, voluntary standard that addresses animal welfare in sheep farms and chain of custody of wool, from certified farms to the final product."[43]

Though the company was unwilling to disclose specific details about the proportion of RWS-certified versus non-certified wool they source, they did state that, "…[T]heir facility adheres to the high standards of both RWS and Good Manufacturing Practices, which advocate not only for animal welfare but also for social welfare." It is well established that RWS-certified wool has been thoroughly audited and the certification has significance within the broader industry. That this company completes the final stages of manufacturing in Canada means we can assume that workers are paid at least minimum wage and that any waste is handled according to Canadian environmental and waste-management standards.

THE BIGGER PICTURE

When it comes to imagining the world I want to live and work in, there are two questions that keep me up at night: Is locally grown and sourced wool always a more sustainable option than wool that is grown and sourced internationally? And, do we have the capacity in Canada to manufacture products, desirable to fibre artists, from materials grown on our own land base? I think tackling the issue of globalized trade and

42 Email exchange with Aamir Dawood, who does not want the company named in the book.

43 Textile Exchange, *Responsible Wool Standard 2.2.* Copyright 2021, https://textileexchange.org/app/uploads/2020/08/RAF-101a-V2.2-Responsible-Wool-Standard.pdf.

production models is complex—and to dismantle or create alternatives is an arduous task. Here are some ideas I propose adding to the conversation, to hopefully inspire some collective "throwing spaghetti at the wall to see what sticks."

Certifications

RWS: Responsible Wool Standard

This certification is specific to wool and focuses on the welfare of sheep. The goal is to ensure traceability from farmer to processor to consumer, to ensure the highest standard of animal welfare, and to reduce the negative impact of livestock agriculture on land health. RWS certification is internationally recognized to distinguish sustainably grown wool from others—this is an important step when we look at the broader wool industry and the environmental and social impacts of our yarn.[44]

GOTS: Global Organic Textile Standard

This certification is applicable to all organic textiles and provides assurance to consumers that every stage of the value chain—growing, harvesting, manufacturing, and labelling—meet organic standards.[45]

OEKO-TEX®

OEKO-TEX is a trademark of the International Association for Research and Testing in the Field of Textile and Leather Ecology. This certification provides confidence in the sustainability of textiles and leather along the supply chain, and ensures the materials are free from harmful substances. This certification also guarantees that the materials used in textiles meet strict safety and environmental standards.[46]

44 Ibid.
45 Learn more about GOTS certification at: https://global-standard.org/.
46 Learn more about OEKO-TEX certification at: https://www.oeko-tex.com/en/.

Let's tackle the second question first. We do not currently have large flocks of Merino sheep (fine wool) in Canada, even though they can, in fact, thrive on our land base, especially in the Prairies where similar breeds like Rambouillet and Targhee manage the grassland range and cold winters well. There is a perception that we lack the processing infrastructure to transform Merino wool into yarn. But if mills like Amtex and Elitespun are already capable of milling imported Merino wool, why is there no industry for Canadian-grown Merino wool?

One barrier is the fact that Canadian sheep require hay six months out of the year during the winter, unlike large Merino-producing nations like Australia that do not deal with our climate. Feeding hay leads to wool that is inherently more dirty with vegetable matter contamination. This contamination requires more work to adequately clean, a process called carbonization, and we do not currently have any processing facilities in Canada that offer this sort of cleaning.

Carbonization

All large-scale commercially produced yarn goes through a process called carbonization to remove any straw, hay, grass, or other vegetation that is normal for sheep that constantly interact with their environment. First, the wool is immersed in sulphuric acid at 6–7 percent weight/volume, before being baked in a large dryer at 120 degrees Celsius. This turns the vegetation into carbon. The wool is then passed through weighted rollers that pulverize the vegetation into dust and through a rotary shaker that removes the dust. Finally, it is passed through a solution of sodium carbonate to neutralize the pH before further processing can happen.[47]

47 W.S. Simpson and G.H. Crawshaw, *Wool: Science and Technology.* Woodhead Publishing Ltd. 2002.

If you have never worked with local farm yarn and haven't had the experience of picking out a tiny piece of hay as you are knitting, it might surprise you that all commercial wool goes through this process to ensure there are no remnants of the pasture or range that the sheep once resided on. The process of carbonization is not done in Canada, but it is always done on wool imported from elsewhere. If we want to increase the amount of fine wool, like Merino, that is grown and manufactured in Canada we need to address the ongoing challenge of "contaminated wool" and how we clean it.

The answer to the first question (is domestic wool always the more sustainable option compared to imported wool?) is determined by what we individually and collectively value in our wool products, and what standards we strive for. If we assess the cost and sustainability of our wool based on its environmental impact and associated emissions, we must consider not only the distance it travels but also the mode of transportation. As highlighted in the chapter on shipping, transporting wool domestically by truck across our vast landmass generates nearly as much emissions as shipping it overseas. Building a stronger regionally grown fine wool industry would help reduce overall transport-related emissions by shortening supply chains and decreasing reliance on long-distance freight.

Prioritizing a domestic industry that doesn't generate an ever-increasing amount of GHG emissions, means focusing on a regional approach to wool production. This approach would start by mapping out geographical resources and sheep population potential. Encouraging large-scale ranches and farms in the rangelands of Alberta, Saskatchewan, and Manitoba to grow larger flocks of fine wool breeds like Rambouillet, Merino, and Targhee would initiate a shift toward more quality fine wool production. At the same time we would need to develop and invest in the proper infrastructure to manufacture fine wool at a scale that is both profitable and socially responsible. If our goal is to process fine wool domestically while maintaining a competitive price point, we must build adequate infrastructure to efficiently process large quantities of fine wool. While mini mills have filled the gap where no processing existed before, they are not large enough to process wool at a competitive price point.[48]

It will not be an easy or swift transition to fill in the current gaps in our manufacturing capacity in Canada. On one hand we have the flexibility and context-specific mini mills that meet the needs of local communities and sheep farmers—but these mills simply don't have the capacity or machinery to adequately process the full Canadian wool clip. On the other hand we have a handful of large-scale mills that have the manufacturing capacity to handle the size of the Canadian wool clip, but not the quality of wool with high veggie matter contamination, the need for large-scale scouring, and comprehensive carding.

Here, I've outlined the crucial missing links needed to advance our industry in a way that has a fully traceable supply chain, and in the Conclusion I will put forward more ideas for how we can shift our wool processing industry in Canada. Ultimately, if we want a new kind of wool industry in Canada that does not follow in the path of our dominant growth economy, it will require collaboration and strategic planning between farmer and processors. It will also require the support and buy-in of the end users and consumer base. This grassroots support will require education and awareness around the true costs of producing and getting our wool to market.

48 Let me explain the scale issue with mills. Most small mills have between 8–15 spindles that spin yarn at one time. If it takes 4 hours to complete a "batch" of yarn we will have a maximum of 15 bobbins of yarn. If we had a larger mill, that has maybe 100 spindles, then at the end of the same 4-hour period we now have 100 bobbins of yarn. This is how we can scale up yarn production and reduce the price per skein.

CHAPTER 4

Processing Wool Internationally

I have given a lecture with the same name as this book, "The True Cost of Wool," more than fifty-five times in the last five years—to guilds, at yarn shops, and at fibre festivals. There is this moment, about thirty minutes into my talk, when I announce that 90 percent of all the wool grown in Canada is actually shipped overseas, while more than 95 percent of the yarn we consume is imported from other places. There is an audible gasp as people look down at the yarn in their hands and the reality of our globalized wool industry becomes real to them.

This export of raw materials and import of finished goods is not specific to the yarn industry—it is the trajectory of much of our globalized production of goods. Canada is a resource-rich country; we harvest raw materials and ship them out of the country for more affordable manufacturing. We have exported raw wool for a long time, first to the United States (remember the Reciprocity Treaty of the 1860s), and then to Britain and other nations overseas. This export trend accelerated in the late 1990s and early 2000s due to the various free trade agreements that Canada signed and was accelerated when China joined the World Trade Organization in 2001. China is now the world's largest manufacturing nation for wool yarn and wool textiles.

Most of the world's wool ends up in China for production[1], so it is no surprise that prior to the 2020 COVID-19 pandemic the majority of Canada's wool was also sold to China.[2] The majority of Canada's wool is bought and sold on the commodities market by the Canadian Cooperative Wool Growers (CCWG).

Founded in 1918 as a producer-based organization to "consolidate and centralize the wool grading services in the Dominion,"[3] the CCWG is the largest wool buyer in Canada and has a long legacy of managing and marketing Canadian wool and working alongside Canadian sheep farmers.

In 1914 the Dominion Department of Agriculture established a special commission to investigate the wool industry. This led to a standardized system for grading wool, and the recommendation that wool be marketed on a cooperative basis. By 1916, there were twenty-six different associations handling the nation's wool. Some of the challenges being addressed were the need for standardized wool grading and ensuring that farmers received the same payment for the same grade of wool. On March 7, 1918, the Canadian Cooperative Wool Growers Ltd. was incorporated.[4]

The "Wool Growers" as the organization is known by most producers today, manages wool in much the same way as when it started. CCWG purchases wool from farmers, then transports it to their central depot at Carleton Place in Ontario. Once there, the wool is sorted, graded, and baled according to its classification and quality. It is then hydraulically packed in large bales and prepared for transport. Farmers are sent a "wool cheque," as many like to call it, along with an evaluation

---

1 "IWTO Market Information Ed.17," International Wool Textile Organization, 2021, https://iwto.org/wp-content/uploads/2022/04/IWTO-Market-Information-Sample-Edition-17.pdf.

2 Eric Bjersgo, "Market Report," *Canadian Cooperative Wool Growers*, Fall 2020, https://wool.ca/page/wool-market-reports.

3 Government of Canada, Dominion Bureau of Statistics, Department of Trade and Commerce, *Textile Industries in Canada, 1941*, 1943, Government of Canada Publications, https://publications.gc.ca/collections/collection_2018/statcan/34-209/CS34-209-1943-eng.pdf.

4 Ibid.

of the total weight of wool that was sent, the quality or classification of that wool, and how much it was worth.

Commercial Wool Rates

For commercial sheep producers who ship and sell their wool to the CCWG, these were approximate rates for one pound of wool in 2022–2023.[5]

Fine wool: $1.25–$2.30
Medium/coarse wool: $0.10–$0.60

It is important to note that sheep producers receive a higher rate of pay for their wool if it is properly skirted (the process of pulling off the dirty or contaminated parts of a fleece after shearing) and sorted. The value will also be higher if the remaining wool has less contamination by veggie matter (grass, leaves, burrs, grains, etc.) because there will be a larger yield. There is responsibility on the individual farmer to prepare their wool in the best way possible to ensure the highest returns—however, when the extra work does not financially pay off, some producers see no value in doing that work. The wool industry is trapped in a cycle where farmers won't reduce vegetable matter contamination or improve wool quality due to low financial returns, while buyers (CCWG) won't pay more for wool that **lacks resale value on the international market.** Waiting for producers to change without financial return or for CCWG to pay without resale value is not working—we need a system shift of revaluing wool.

5 Rates taken from the CCWG "Wool Grading Statement 2022."

The CCWG then sells Canadian wool to the international commodities market. According to their mandate, the CCWG aims "…to be a producer cooperative endeavouring to market wool at the top price it deserves while operating the business efficiently to maximize returns to the wool producer."[6] Despite the incredibly low prices that CCWG pays to producers for their wool (between $0.10 and $2.30 per pound), they do not set global wool prices. The global wool market has shifted as it has been subject to many of the political, environmental, and health-related changes the world has faced in the last twenty to fifty years—things like free trade agreements, acts of terrorism and war, recessions, trade conflicts between global superpowers, SARS, COVID-19, growing demand for synthetic fibres, oil prices, and more. Reading the annual wool market reports for the last thirty years provided by CCWG, we can see that the wool industry has had so many ups and downs it can be hard to pinpoint one specific event, trend, organization, or institution that is responsible for the current situation.

Because CCWG did not respond to my request for details about where Canadian wool goes after they purchase it, I will work under the assumption that the wool is transported via transport truck to the United States, or shipped via the Port of Montréal to various international destinations, though they may also use the Port of Vancouver. The only information I received from CCWG regarding shipping methods was in an email that stated, "Since CCWG bears the transport cost of moving wool from its depots, the aim is to ship wool in an economical fashion."[7]

We do know for certain that when the CCWG markets Canadian wool to China, the United States, and other international buyers, there is very little traceability around what happens to the wool or where it ends up once it leaves the port. The further removed we are from the production of our goods and materials, the less information we have about how it is produced—including the environmental implications,

6 Canadian Cooperative Wool Growers Website, https://www.wool.ca/page/about-us (Accessed September 2024).

7 Email correspondence with the CCWG. November 30, 2024.

effluent treatment, carbon emissions, labour standards and wages paid, and end-product waste disposal. But this lack of transparency is not inevitable in cross-border production. When we have local production, we can trust that at the very least the environmental and labour standards determined by Canada are being adhered to.

For many years, Canadian wool has been increasingly undervalued. Lack of competition or innovation within the large-scale Canadian wool-handling industry has led to stagnation and a resignation among farmers and other stakeholders that Canadian wool is at the mercy of international commodity prices. The CCWG has seemed slow to invest in local solutions for Canadian wool. CCWG, as a cooperative business, has, however, been able to maintain and even grow the business through their livestock supply retail division. The company has seen profits and financial growth even while the wool business has declined. Perhaps it is not my place to suggest what a company that I am only recently a shareholder of, should do with their profits.[8] But with such a long-standing tradition of working in the best interests of sheep farmers, I think CCWG should re-invest in building up the domestic wool industry and financing infrastructure that is well suited to the type of wool we grow. They are in an excellent position to be innovators in wool and would ingratiate themselves with sheep farmers and fibre artists alike, a major contrast to the unenthusiastic relationship that exists for many now.

In our effort to "Follow the Wool," this is the end of the story for the majority of Canadian wool. Despite numerous attempts to obtain more detailed information from CCWG about what happens to Canadian wool beyond their warehouses there was no information provided. However, CCWG is a producer-owned cooperative and they can change their business operations at the request of the majority shareholders. They are well positioned to support a domestic wool industry and to invest in wool processing and product development. In the years that I have been engaged in the wool industry I have observed increased

8 Anyone can become a shareholder in the CCWG for $50. Being a wool producer is not a prerequisite. https://wool.ca/page/become-a-shareholder

frustration and pressure from CCWG shareholders to push the company to investigate local options. It appears that this pressure has had an impact, and CCWG seems more open to supporting a domestic wool-processing industry through creation of a new innovation prize, and research into the viability of a wool pellet operation.[9]

FOLLOW THE WOOL

Commercial Wool Tally

Let's look at the final true cost of wool raised by commercial producers:[10]

Sheep costs for one sheep for one year: $200.00
Shearing: $7.00
TOTAL: $207.00

Total wool yield per sheep: 8 lb.
Price per pound for medium domestic wool: $0.50/lb.
TOTAL: $4.00

When we look at these numbers, we can see that it is not financially feasible for a sheep farm to exist solely on the sale of wool as a commodity (selling wool commercially through CCWG at current commodity prices and given the medium and coarse quality of the majority of Canadian commercial wool). In fact, it often costs more to shear a sheep than the wool is worth once it's sold commercially. In this context, the wool is worth less than 5 percent of the total revenue

9 Wool pellets are a soil amendment that transforms poor quality or dirty wool into pellet fertilizer. This value-added product made from wool 'waste' is growing in popularity among small producers in Canada since 2020.

10 This final tally was based on an average cost projection from the farmers surveyed for Chapter 1 and wool evaluated as a medium quality domestic wool.

generated from each sheep.[11] For a sheep farming operation in Canada to be economically viable it must also have revenue from each sheep through meat, breeding stock, and lambs produced. Until it is worth more, wool will continue to be an expense rather than a contributor to net revenue.

When faced with this reality, it is not surprising that many sheep farmers are not interested in improving their wool quality unless a significant increase in pay were to be presented. I do not believe this is an impossible scenario—in fact, I have worked with many commercial producers to improve some simple aspects of their operations to increase the quality and cleanliness of their wool to a point where they can sell it to small mills or handspinners for closer to 10 to 15 percent (or higher) of the total revenue generated from meat, lambs, breeding stock, and wool. So, while improvement of wool quality and selling direct to fibre artists can impact the bottom line of some commercial sheep farmers, it is not the kind of revolutionary change we need, to have the greatest impact on the sheep industry as a whole. The kind of revolutionary change we need requires larger-scale infrastructure for processing wool domestically and a product development that utilizes the wool that is produced currently (medium and coarse wool) with a plan to grow the fine wool industry in the future.

It seems impossible that there could be a price increase significant enough for farmers to see the value in wool when it's sold into the global commodities market. As demonstrated above, the global wool market for medium/strong wool has very low value creation, and farmers are certainly not reaping the benefit of the end-product pricing. The CCWG (or any other large wool buyer/seller) is handcuffed by the global market and lacks influence to change the system. If wool industry institutions and companies played a more active role in developing the wool value chain in Canada, we would have a traceable system and

11 To calculate this number I took the total revenue generated by one ewe, including market lamb prices and the revenue from breeding stock, and combined wool sales, then divided the wool revenue by the total to determine the percentage of revenue for wool only. Market lamb and breeding stock prices were an average of the numbers provided by the farmers I interviewed.

the agency to make changes and develop products that are suited to our wool. It would involve a large initial capital investment to build the necessary processing infrastructure to develop wool products domestically, but in the long run this would increase the overall price per pound for farmers, it would provide jobs for Canadian workers, and it would ensure greater resilience in the face of the kind of external factors that have wreaked havoc on the global wool industry for decades.

CHAPTER 5

Marketing Wool

My first eight sheep were sheared in the spring of 2017. I spent hours skirting those fleeces and making sure they were prepped and ready for processing. After the embarrassment of realizing there weren't any mills in Manitoba, I found Chris Armbrust at Dakota Fiber Mill in North Dakota, and booked a time to send her my wool. By the time the wool was processed and ready to pick up, I was practically buzzing with anticipation.

I had no idea what to expect, but when I saw the bags of spun, plied, skeined and ready-to-go yarn I was actually slightly disappointed. Not in the quality of the work by Chris and her team—she had done a beautiful job of transforming my wool into yarn—I just couldn't believe there was so little yarn to take home. I had brought eight large garbage bags full of wool to the mill—it took up the entire trunk and part of the back seat of my car—but when I picked it up as yarn, it fit neatly in one large box.

Chris could probably see the shock in my eyes. (Now that I have processed dozens of farmers' wool I, too, have witnessed that shock and disappointment when they see the final yarn tally.) She immediately explained that in the scouring and processing, you lose up to 50 percent of the weight of the wool you started out with. I let my surprise go and began the exciting process of tearing open every bag to fondle each individual skein. I couldn't wait to get my first ever homegrown yarn out into the world.

The price of the yarn can obviously vary depending on the business, location, and capacity of the producer. Once a fibre farmer has their yarn back from the mill, they face the task and added cost of dyeing it, labelling it, taking photos of it, loading it onto a website with a description or taking it to a wool show, marketing it to customers, packing and shipping it. This work can be extra challenging for producers who are still managing a farm with sheep that need to be tended to daily. It would be quite a challenge to generalize the costs for every local yarn producer across the country, but here is an example of some of the extra costs of bringing yarn to market.

FOLLOW THE WOOL

Marketing Costs for 1 Skein of Yarn Sold at a Canadian Yarn Show

Naturally dyed = $4.00
Yarn label = $0.24
Booth fee = $1.05
Travel costs = $0.87
TOTAL = $6.17

As a small-scale fibre farmer travelling to large knit shows I try to have between 400 and 600 unique items to sell. To get the price per skein for booth costs and travel costs I took the total amount and divided it by 500. These numbers would vary depending on how much the booth fee, travel costs and number of products available for sale are, but I'm aiming to give you an estimate.

Labour Costs

You may have noticed that I haven't included any wages for the producer up to this point. Some farmers may be more disciplined than I am and take a monthly or weekly salary from the products they sell on the farm, but for many of us it is more like a mad scramble to make ends meet. Yarn and wool can be a very seasonal business and the sales ebb and flow as the weather changes. I have not factored in an hourly or weekly wage rate for myself or for my husband because I think if we calculated a living wage for farmers it would not capture some of the non-monetary value we get from the life we have chosen. That being said, farmers should absolutely be adequately compensated for their work.

When the doors open at Knit City, Twist Festival, Landmade, WoolStock, Fibrations, Prairie/PEI/Manitoba Fibre Festival, or at any of the other big and little shows across the country, producers wait with anticipation, hope, and sometimes even a little desperation to go home with empty bags and enough cash flow to keep their operation going. This is where, as farmers, our wool journey ends and yours, as the consumer, begins. Here is a breakdown of the total true cost to get this yarn to your hands:

FOLLOW THE WOOL

Final Tally for Fibre Farm Yarn

Here is the True Cost of Canadian locally grown and produced yarn in all three scenarios (see page 34).

Scenario 1

Sheep costs for a year: $15.07
Shearing cost: $0.60
Shipping costs to mill: $0.79
Manufacturing costs: $14.73
Marketing costs: $6.17

True cost per skein: $37.36
True value retail cost per skein (industry standard is to double the price of the production cost): $74.72

Scenario 2

Sheep costs for a year: $13.07
Shearing cost: $0.36
Shipping costs to mill: $1.03
Manufacturing costs: $14.74
Marketing costs: $6.17

True cost per skein: $35.37
True value retail cost per skein: $70.74

Scenario 3

Sheep costs for a year: $8.80
Shearing cost: $0.37
Shipping costs to mill: $0.34
Manufacturing costs: $14.32
Marketing costs: $6.17

True cost per skein: $30.00
True value retail cost per skein: $60.00

I can imagine that final number is a bit staggering to you. It was to me the first time I added up the true cost of production. Now that we have an accurate cost to produce quality farm yarn in Canada, we can do some comparison to the other wool that is available at fibre art shows, in stores, or online. Of course there is a range of prices for wool yarn, but generally the price point ranges from $10/100 g skeins at the lowest end, up to $50/100 g skein on the higher end. If the price for local wool yarn is between $71 and $86, that limits the number of customers that can afford it.

In my experience as a sheep farmer, yarn producer and advocate of local wool manufacturing I see a growing trend toward local wool with the transparent value chain that I've discussed in this book. There are more customers shopping locally and trying to connect with their local sheep producers and mills. There is often a sentiment that it is hard to find local wool in yarn shops or at knit shows and that local wool needs to be more readily available at a price point that is affordable.

I would like to outline some of the barriers and opportunities facing fibre farmers in accessing bigger markets for local wool.

Wholesale to Local Yarn Stores

Many local yarn store (LYS) owners have reached out, asking to stock my yarn, or to provide them with the names of farmers who offer wholesale rates on their locally grown and -milled yarns. Although this would be the optimal channel for getting more Canadian wool into the hands of fibre artists, it isn't a feasible option for me.

For many yarn companies, the wholesale cost of yarn is 50 percent of the retail cost. This profit margin allows LYSs to keep the lights on, pay their staff, bring in even more yarn, and provide all the benefits of a brick-and-mortar yarn shop.

For them to stock their shelves with small-batch, locally grown yarn and wool, the numbers just don't make sense: If the cost of producing a skein of yarn is $35.00, the fibre farmer has to sell it to the LYS with at least a small profit margin on top. Let's pretend that they wholesale it for $42.00 (that is only $7.00 profit on each skein). Then, the LYS has to turn around and generate their own profit in order to manage their store. If they follow the industry standard of doubling their wholesale cost, they would price the yarn at $84.00, and it becomes financially inaccessible to many crafters. The LYS could only mark the yarn up by 15–20 percent, and then they would be limited in how much local wool they could carry because they would not be generating enough revenue from farm yarns to stay viable.

The major exception to this scenario are the two large Canadian mills—Briggs & Little and Custom Woolen Mills—which can operate at a larger-scale and thus produce a greater number of skeins of yarn for a lower production cost. Let me explain: A hypothetical batch of yarn takes the same four hours to spin on the bobbin of an industrial spinning machine as it does on the bobbin of a smaller spinning machine, but you will get a lot more skeins out of a 100-bobbin spinner than a twelve-bobbin spinner. The end result is much more yarn from the bigger mill.

The ability to scale up a spinning operation comes down to the size of the infrastructure. Sadly, there are not many large-scale mills in Canada, and there are not many people with the skills or knowledge to run them.

VALUE-ADDED PRODUCTS

Many fibre farmers have pursued other product options for their wool (in addition to or instead of yarn) including wool pillows, duvets, blankets, batting and roving, felt, felted clothing, or crafting kits. These are all creative ways to transform wool into value-added products. Many across the country have tapped into spinning or felting groups to direct-market their roving and batting. Some farmers skip the spinning step and turn their wool batting into pillows and duvets. Many have invested in small-scale felting machines or worked with an

Ontario-based felting mill to create felted wool insoles, dish scrubbies, pot holders, carpets, acoustic panels, or clothing. Others have partnered with weavers or invested in weaving equipment to make blankets, or fabric for clothing.

All of these unique ideas offer a local wool product to a distinct market and create opportunities for growth in the industry. I think we have only scratched the surface of what products can be developed and what markets can be explored.

Every new product that is developed requires processing and an extra step along the value chain. These steps all come with a price tag and more labour—the key will be to collaborate with others to share the cost of these steps through larger-batch processing or to invest in more efficient or larger machinery to cut down on production costs by creating an economy of scale.

Challenges to Direct Marketing

In the winter of 2020, I conducted a small research project to capture data on some of what was happening with smaller-scale fibre farms.[1] I wanted to know who was growing wool and selling it directly to consumers. These numbers are not captured in Statistics Canada studies, but they are important for a well-rounded understanding of the industry. I wanted to know where these farms were located, who was running them, and what were the big challenges they were facing.

My assumption as I went into the research was that lack of access to processing would be the number one barrier facing fibre farmers. And while access to processing turned out to be the second highest barrier, the number one barrier was marketing. In follow-up interviews, many sheep farmers stated that they are so busy feeding sheep and tending to their other livestock responsibilities, plus getting the wool processed and turned into products, that they simply don't have the time, energy, or sometimes even the knowledge to effectively bring their wool to market.

1 Survey results and methodology can be found on my website: https://www.longwayhomestead.com/survey-resultus.

Despite the growing interest among consumers to "buy local," and renewed engagement with local sheep breeds, it is challenging for fibre farmers and small mills to adequately perform their day-to-day operations and stay on top of the constantly shifting market landscape. I often encourage eager crafters or fibre artists to take the initiative and reach out to farmers, to offer support on shearing day, or even to just drive out to the farm to pick up an order of yarn or fibre.

The numbers and examples I've laid out in this book demonstrate that fibre farmers do not have a huge profit margin on growing wool and selling value-added yarn or other products. Many have a diversified farm, so they not only sell yarn but also do agri-tourism or sell meat or breeding stock or host workshops. Many work part-time or full-time jobs off the farm while also managing the farm business. Many of us do it for other reasons as well—love of land and animal, the ability to live in a rural community, and/or to have a higher degree of connection to our food and fibre. Or perhaps we have found a good way to subsidize our love of good quality wool, or to keep rare breeds of sheep from going extinct. Whatever the reasons, the decision to farm is often driven by more than just profit. When a fibre artist approaches my booth at a knitting show and gently grasps a skein of yarn and lifts it to their nose and sighs, "I love this—it smells like sheep!" it validates the hard work I do and it connects me to the broader community.

Validation is important—but so is paying the hay bill for those sheep every year. As producers of wool we need more than a pat on the back for doing a good job. It is easy for fibre farmers and their wool to get lost in the sea of beautiful, flashy neon yarn, or rich, tonal superwash yarn. The base price for farm yarn is higher than that of commercially available yarn because farmers/local wool producers bear the entire true cost. This shrinks our profit margin and limits the opportunity for business growth and investment into scaled production models. Commercial yarn companies don't bear this burden of true costs because they are removed from the various stages of wool production. They can purchase wool at low prices, manufacture in facilities with cheap labour, and leverage their purchasing power to ensure those prices stay low.

Small-scale fibre farmers not only have to advocate for ourselves and our products at the manufacturing level, but also at the consumer level. The onus is on us and on Canadian wool-based yarn companies to educate consumers about why our products are different from commercially produced products, and why they should care. Perhaps this is just the cost of doing business—but I would put forward the opinion that other commercial yarn companies do not pay the true costs that their supply chain actually incurs. We need to put some of the responsibility back on the large companies to share the full traceability of their products and to be transparent about those true costs, and I would suggest that they also need to take responsibility for paying for those true costs.

There is a growing movement of people that are focused on supporting local wool economies and seeking out wool that has a fully traceable value chain. As fibre farmers, we have so much gratitude for these champions of Canadian wool—but we need more! More attention and amplification in knitting and fibre arts spaces and shops, more access to fibre shows and knitwear designers that want to feature farm yarn, more education and information widely available about the unique characteristics of working with farm yarn and the challenges of growing, manufacturing and marketing our local wool supply chains.

Industry in general, and the agricultural industry specifically, receives millions of dollars in subsidies and infrastructure investment every year from both the federal and provincial governments.[2] These publicly funded subsidies come in the form of insurance programs or direct payments to encourage production, compensate for low market returns, or stabilize income.[3] If we use public funds to support existing

---

2 Government of Canada, *Direct payments to agriculture producers (× 1,000)*, Table: 32-10-0106-01, Statistics Canada, Release date: February 28, 2025, https://www150.statcan.gc.ca/t1/tbl1/en/tv.action?pid=3210010601.

3 Government of Canada, *Direct Payments to Agriculture Producers*, Statistics Canada, https://www23.statcan.gc.ca/imdb/p2SV.pl?Function=getSurvey&SDDS=5229.

agricultural production, then surely we could collectively push for subsidies or new investment in the infrastructure needed to make Canadian-grown and -manufactured wool competitive in our local markets. We need a new system of wool production that is subsidized FOR farmers rather than subsidized BY farmers, and as consumers of wool and yarn we all have a role to play in advocating for this new system.

CHAPTER 6

Changing the Canadian Wool Industry, One Sheep at a Time

In the fall of 2016, I got a phone call from a local sheep farmer who wanted to know if I'd like to pick up three large bags of wool that they were going to throw out because it wasn't worth anything to them. I had only been farming sheep for four months at that time, and starting a mill was still just a passing thought. I felt compelled to save this wool from the landfill, so the next day I drove my truck to their farm and picked up about 600 pounds of it. To be honest, I had no idea if the wool was good quality (it wasn't) and even less of an idea of what it would become (it ended up as garden mulch). Over the next two years these three giant bags of wool were moved around the farm from the garage to the greenhouse to the shop and finally to a storage shipping container as I figured out how to process wool. They were a constant reminder to me of how big the challenges within the wool industry are—but they were also a motivating reminder that all I had to do was figure out how to deal with one bag of wool at a time.

That wool was what I cut my teeth on in this industry, and as the last of it is currently breaking down in my garden and providing valuable nutrients to the soil ecosystem. I am steadfast in my belief that we don't need to have all the answers to solve these big problems—we just have to be open to possibility and willing to take a risk and try something new.

I want to spend some time looking at the folks across the country who are doing the hard work of revaluing Canadian wool. These people are not waiting around for someone else to change the industry; they are doing it themselves—in the sheep barn, in the mills, at the wool shows, in government, and in fashion schools. I have always believed there isn't one big answer to the challenges we face, but many answers, and I see that every time I hear what a farmer or a community is doing.

I must start by acknowledging the hard work that many have been doing for decades—the fact that we can stand on their shoulders to evoke even more change is significant. One of my mentors, Gerry Oliver—an outstanding sheep farmer who has worked tirelessly for more than twenty-five years promoting rare breeds, educating others on shepherding and fibre management, and advocating for wool—said to me recently, "I've been saying the same thing for many years and it feels like things are perhaps starting to change."

FARMING—IMPROVING WOOL QUALITY AND MAKING IT PAY OFF

We can't have a wool industry without farmers who continue to raise sheep. There is a growing number of sheep farmers that are focusing on wool quality, direct marketing to customers, breed conservation, and animal welfare. Many of these farmers have figured out how to navigate the wool supply chain and have carved out a niche for their products in their own community. They have developed unique products like felted insoles, blankets, wool socks, wool pellets, and more. I encourage you to seek out these farmers, purchase their products and invest in our local fibre economy.

As we grow the Canadian wool industry we will have to confront the consistent challenge—one that we have faced since the early 1900s—of our wool being predominantly more coarse and not fine quality like the Merino that has become the global standard. We need more products that are well suited to our Canadian wool clip, such as carpets, bedding, insulation, or acoustic panelling. There are many entrepreneurial farmers who have been working on this, and one example is Romy and Ryan Schill, who operate Circle R Lamb Farm and Revolution Wool Company in Ontario. They have been running a large-scale sheep farm for more than fifteen years, with a focus on producing high-quality lamb. Romy was not satisfied with the commercial per-kilogram rate she received for her wool and decided to navigate the supply chain and develop value-added wool products. She works with numerous mills throughout Ontario and has a growing business that focuses on wool bedding (pillows and duvets), felted products, blankets, and more. In 2024 Romy added an industrial duvet sewing machine to her operation and solidified her position in the industry as someone who produces high-quality wool bedding. Taking advantage of some of the remaining larger-scale wool-processing infrastructure in Ontario, Romy has been able to develop a value chain for her wool that is efficient and inspiring. She has demonstrated what is possible for large-scale flocks of "meat sheep" when the wool is valued and transformed into products best suited to that type of wool.

WOOL BROKERS—THINKING OUTSIDE THE BOX

Of course, not every sheep farmer is going to have the time, desire, or ability to establish a value-added wool-based business in addition to all the farm (or off-farm) work they are already doing. This is where I see great opportunity for a new wave of wool brokers to emerge across the country. For years, Canadian sheep farmers have had to rely on the CCWG to buy and sell their wool—it was the only option around. I have witnessed how a new type of collaboration and partnership between sheep producer and wool broker can result in higher values for the wool and a more streamlined process for manufacturing. This collaborative approach has the potential to revolutionize the industry. Decentralizing the acquisition of wool, and the production of wool products, results in better distribution of financial risk, decreased GHG emissions from centralized shipping, and increased resilience and flexibility in navigating supply chains.

I have seen some excellent small-scale examples of this by innovative entrepreneurs who wanted to work directly with farm wool but didn't want to start a sheep farm. Catherine Knuttson of Small Bird Workshop and Brittany Vogt of Crux Fibres are two excellent examples of this kind of small-scale wool advocacy. In their own unique ways they have built relationships with sheep farmers in their respective regions. Catherine attends shearing days and hand-picks many fleeces based on her extensive knowledge of sheep breeds and her relationship with the farmers themselves. She focuses on high-quality, breed-specific fleeces and prioritizes paying sheep farmers what the fleeces were worth. She ships fleeces to various small mills and has them transformed into yarn and spinning fibre, and then markets them under her label to her customers. She is known in the community as someone who cares deeply for farmers and the quality of the wool, who understands the small-scale wool value chain and who, of course, makes beautiful products.

Brittany lives and works in the Yukon and doesn't have access to local sheep farmers, but that didn't stop her from developing her own take on local wool called Low Mileage Wool. Brittany works with sheep farmers in Alberta to hand-pick fleeces that are then processed at mills throughout Alberta and Ontario. She produces beautiful yarn and fibre products that tell a story and are deeply connected to the farms they come from and the shepherds doing the work. Brittany will often partner with these farmers to attend knitting and fibre shows and to further foster the connection between knitter and farmer. In her products and marketing and through her presence at wool shows, Brittany truly advocates for low-impact wool and greater connection to sheep farms.

These are just two examples of innovative approaches to regional wool systems that could be developed further. There are many more examples of small-scale wool brokers in every region of the country—and with all the wool that is wasted every year there is room for even more of these kinds of ingenious business models. These businesses are proof you don't need to own a farm or wool mill to drive wool through the value chain.

The last example I'll highlight, of new and resourceful approaches to the wool value chain, is the work of Jane Underhill and her business, J. Underhill, Canadian Wool Agency. You may recognize Jane's name because she was the lead researcher for the Campaign for Wool's Wool Plan, released in 2021. Jane grew up in Moncton, New Brunswick, and had a farm-adjacent upbringing that has helped her connect with and build relationships and trust with sheep farmers through her work. After many years running an unrelated business, Jane became aware of the problems within the wool industry in Canada and was determined to make a change. Her business has three main functions: 1) white label manufacturing, 2) research and consulting services, and 3) farm services. Jane works with stakeholders along the wool value chain "...to develop solutions that enhance the quality, performance, and market competitiveness of Canadian wool."[1]

Jane works with larger quantities of wool than some of the smaller wool brokers—in two years she moved about fifty tonnes of Canadian wool through the supply chain. Jane values the network and connections she has with people who are invested in the wool industry and she works to bring people together to find solutions to the challenges facing the industry. Jane also prioritizes a reduced carbon footprint in her work and has developed a value chain that exists within a "... two-hour radius." This also cuts down on manufacturing time and travel costs.

Catherine, Brittany and Jane's, bold, adaptive approaches prove the potential for new solutions to our under-utilized wool.

These examples of wool innovation require access to production and manufacturing. There are some immediate barriers to growing our domestic wool production capacity in Canada. Cost and access to funding, equipment, machinery, and skilled, affordable labour are the big ones that are mentioned in most conversations between farmers, brokers, mill owners, and other industry players.

1 Correspondence with Jane Underhill, January 2025.

CHALLENGES FOR INDUSTRY GROWTH...

When my husband, Luke, and I decided we were going to open a wool mill on our farm in 2018, we spent considerable time researching the viability and opportunity for a mill—if we were going to borrow $300,000 to start a mill we had to know that it could work. I emailed or called many other mill owners, and most of them were hesitant to talk about their own experience. A few spoke of the long hours and hard work for very little pay—we even found a financial viability study in our region that stated it was impossible to make a wool mill profitable. There were a lot of words of warning for us as we explored the financial challenges of small-scale wool processing.

We went ahead with the plan anyway, and six years later I can honestly say that it is one of the hardest things I've ever done. I understand why there aren't more domestic wool production start-ups. The first challenge was financing—where could we access enough capital to start a wool mill? We ran a successful Kickstarter crowdfunding campaign and raised about $12,000 toward our mill. The generosity, support, and confidence that our community has in us is a constant motivator. We went to traditional banks and to the Canadian agriculture-focused bank, Farm Credit Canada, and the answer we got was the same: "No, you don't have any financials to prove that this will work or that you can do it, and you don't have any collateral to justify a loan for this amount of money."

I finally found the Women's Enterprise Centre of Manitoba and they were excited, passionate, and willing to partner with me on this project. We literally "bet the farm," putting up our house and farm as collateral for the mill, and then we jumped in head first to make it work. My experience in pursuing funding for our mill represents a similar barrier for many in opening up domestic processing facilities. It is costly to purchase used or new equipment, to find and secure a location, to do the necessary upgrades to accommodate wool production—and to secure funding to pay for it all.

New mills in need of start-up capital have a few options available to them: small business loans from banks, donations and/or loans from their community, grants from government or private foundations. Each individual mill or business has to decide their level of risk, access, or ability to utilize funding from these different streams.

If a potential new mill owner is considering a bank loan, the possibility of a higher interest rate due to the risk factor can add extra pressure to immediately produce quality products and generate profit. For those who have worked with wool, you know that it is a dynamic material that requires skill and experience to turn into a consistent quality product—learning how to operate a wool mill requires a significant amount of time. The balance of having to make loan payments while learning new machinery can be an overwhelming step for even the most confident new business owners.

There are very few agriculture grants available federally or provincially for new start-ups in the wool or agri-product manufacturing industry. Most of the grant and subsidy money available in Canada currently is for existing large-scale agri-food businesses and requires applicants to provide three years of financials—something that new start-ups don't have. Additionally, because wool is not a well-defined agricultural product in Canada, any application for grants or subsidy money requires the applicant to defend and/or prove that wool production is beneficial to the Canadian agricultural economy.

Sustainable Canada Agricultural Partnership (SCAP) funding, introduced in 2023, has the potential to support the creation of new wool manufacturing. The SCAP agreement is a "five-year, $3.5 billion investment by Canada's federal, provincial and territorial governments that supports Canada's agri-food and agri-products sectors."[2] But it is limited to a cost-share agreement of only 25 percent support from the government and is challenging for new manufacturing businesses to access.

There is limited financial support for those who are serious about starting a wool processing business in Canada—they must be driven and determined with a passion for the wool industry and small-scale

2 Government of Canada, Sustainable Canadian Agricultural Partnership, Agriculture and Agri-Food Canada, https://agriculture.canada.ca/en/department/initiatives/sustainable-canadian-agricultural-partnership (Accessed September 2024).

production to see it through. The return on investment for a wool processing mill in Canada is slow and steady. We need a patient capital model; a model where sustainable growth and community impact is prioritized alongside financial return. We also need to push every level of government to invest in the wool industry, because it makes financial sense to build up industry that provides jobs and reliable supply chains. Our recent collective experiences of a pandemic and a trade war that compromised globalized supply chains, have shown how important it is to decrease our reliance on cheap offshore labour and carbon-emitting transport.

...AND OPPORTUNITIES FOR INDUSTRY GROWTH

I have a serious amount of respect and gratitude toward anyone who starts a mill in the current economic climate. But visiting Wave Weir and her new mill in Seguin, Ontario, left me inspired and hopeful for the future of small-scale processing in Canada. I get the sense that Wave is the kind of person who is only more motivated when someone tells her an idea is impossible.

The Wave Fibre Mill is a semi-worsted spinning and weaving mill that is the embodiment of Wave's vision for the future of textiles and sustainability. Her website makes it clear what motivates her: "Working from the basic philosophy that economies must support a thriving community while doing no harm, we engage a local workforce and area artisans, and look for local, pure, unadulterated, bio-degradable materials to create each garment."[3] Wave has worked tirelessly to build the mill, workforce, and community to create clothing and yarn that reflects her values for high-quality wool products, expertly designed garments, and a truly regenerative life cycle for the resources used in the process. Wave not only processes raw wool into finished yarn but she also has numerous weaving looms that produce quality fabric that is then fashioned into functional and beautiful apparel. Wave partners

3 Wave Fibre Mill, https://wavefibremill.com/ (Accessed August 2024).

with designers and clothing boutiques to bring 100 percent Canadian cloth and apparel into fashion houses and runways. She not only wants to create connections with farmers and processing, but to remind people that clothing is an agricultural product that you can see, wear, and feel inspired by.

Building a Little Army of Wool Activists

I can't talk about the work of revaluing Canadian wool and building a new type of industry without focusing on all the dedicated consumers and end users and knitters and crafters and wool-lovers! I've attended so many wool and knitting shows across the country, and at every event I'm greeted by their overwhelming excitement and contagious enthusiasm for Canadian wool. Fibre artists who absolutely love supporting the farmers, mills, and small businesses focused on local wool are willing to support them with their money, their encouragement, and their voices.

In 2020 I wanted to be a part of building a broader movement of fibre artists that were passionate about Canadian wool and wanted to make personal and systemic changes within the industry. I felt at that time that it needed to be something separate from my business—from my paid work—and this is how Canadian Wool was born (see canadianwool.org).

It simply started as a website to share research and ideas, and then it became an online space where consumers could be connected to wool farmers by accessing a database to look up the sheep farmers who sell wool—raw or finished goods—in a specific region. This imperfectly executed idea actually took off, and more and more knitters started reaching out—wanting more, wanting to keep the conversation going beyond the knitting shows, wanting to take an active role. I started hosting monthly virtual "coffee houses," where an individual involved in the wool industry—a farmer, a shearer, a wool broker, a mill owner—would come and share what they are doing with wool. Then we would openly discuss the challenges and opportunities in the industry and brainstorm ways that "just a knitter" could participate in trying to address them.

The resulting Canadian Wool Collective became an official national non-profit organization in the summer of 2023, and has since launched a Canadian Wool Certification program. This program "is an initiative to create a nationally recognized logo representing wool and value-added wool products that were grown and manufactured in Canada,"[4] with an explicit goal to support Canadian sheep farmers and manufacturers and to advocate for increased transparency and traceability within the industry while educating consumers about the benefits of wool. The collective launched the certification program in the summer of 2024 and continues to be a strong voice of engaged fibre artists, mill owners and farmers, who are taking a grassroots approach to revolutionizing the wool industry—one knitter at a time.

4 https://www.canadianwool.org/certification

Conclusion

WOOL CAN CHANGE THE WORLD

"Wool can change the world" is something I often say during tours of our farm. Most people look at me slightly amused but mostly skeptical.

"Seriously," I continue, "sheep are superheroes. They have the ability to transform grass and sunlight into meat, milk and wool. They are the ultimate upcyclers—they take nutrients from plants (that we can't digest) and turn it into a keratin-based protein that we can spin and knit into clothing. Not only that, when sheep are well managed on our land base they can actually encourage greater carbon sequestration in our soils.

"They transform inedible plants into clothing that keeps us warm or cool (wool is hygroscopic or temperature regulating, by the way). And through the process of doing this they also help our soil ecosystem capture excess carbon from the atmosphere and store it in the soil to feed microbes that will then help the grasses grow again, which will lead to more wool sweaters."

If the farm visitors aren't all obviously wowed by this truth bomb then my llama, Leo, will sense that I'm losing the crowd and walk over to batt his eyelashes until everyone claps.

But I don't need a cute and dorky llama to reinforce the power that sheep and wool have to truly change the world.

Today, 64 percent of all our textiles are synthetic. Derived from fossil fuels—petroleum-based chemicals and petrochemicals—these textiles are actually de-sequestering carbon that has been stored for millions of years in the ground.[1] Even with recycled plastic bottle t-shirts, our synthetic clothing (and yarn) is a one-way pipeline of petroleum out of the ground and into the landfill, where it takes anywhere from 40 to 200 years to decompose. Meanwhile, wool only accounts for about 4.3 percent of all the textiles that we use. Wool is 50 percent carbon and has the potential to not only store carbon from the atmosphere, but to recycle (at the end of its useful life) that carbon back into the soil—in less than *six months.*[2]

We need to divest from the synthetic-textile and yarn industry and build up a fibre economy of wool and other sustainable or regenerative textiles. And we need to take care not to trade one material for another without also choosing models of production and value chains that represent the true cost of these materials.

The further removed we are from the value chains of our wool and textiles the less we can know about who is making the decisions and what factors influence their decision-making. In a globalized capitalistic system, profit is a key driver. This means that we are often stuck in a cycle of trade-offs: If we want a lower price, then we have to pay lower wages. If we want to produce more and do it faster, then we have to increase our reliance on fossil fuels or sacrifice quality craftsmanship for the sake of efficiency.

> *When we purchase a "cheap" ball of wool yarn, the cost of production wasn't reduced—it just means someone or something else subsidized the cost. Perhaps it was the farmer, or the mill worker, or the rivers and lakes or the climate—we can't know these answers if we don't have a clear view of the supply chain our wool travels from beginning to end.*

1 "A new textiles economy: Redesigning fashion's future," *Ellen MacArthur Foundation*, November 27, 2017, https://www.ellenmacarthurfoundation.org/a-new-textiles-economy.

2 "Materials Market Report," *Textile Exchange*, September 26, 2024.

In our mill, we can tell you the name of the aquifer that our water for scouring comes from. Then we can tell you what happens to the waste water. We encourage people to visit our farm and mill and to observe the impact of our business on the environment and animals. Perhaps this is an extreme form of traceability, but imagine if you were able to access information from every yarn company about the process that their yarn went through. We can draw a line from the sheep that grew our wool, to the hands that sheared it, to how it was transported and how it was processed in our mill. We can quantify the waste that our processing produces and how we deal with it. I believe this traceability and transparency have resulted in a loyal customer base who feel more connected and invested in the entire wool industry because of it.

The COVID-19 pandemic revealed the fragility of our supply chains and the potential disaster of our current globalized systems. During that time, many turned to knitting or fibre arts for comfort, distraction, or even for a way to take control of their own clothing. This pause gave us a chance to think about a system where bigger isn't better; where maximizing production, reducing costs, and increasing profit aren't the most important priorities. We had a glimpse of a world where we can build regional, resilient, and diverse food and fibre systems that focus on local supply chains, community-based economic development, meaningful employment, and ecological regeneration. As life gets back to "business as usual" I hope we don't forget that moment in time where we all imagined a different world was possible.

WHAT IS THE SOLUTION?

It can be a bit of a heavy burden to put forward solutions to the problems that face us in the wool industry today. My first inclination was to list the individual choices we can all make that have a big impact on the day-to-day lives of fibre farmers, but then I was reminded of an article I read recently by journalist Mark Kaufman where he revealed:

> *"British Petroleum, the second-largest non-state-owned oil company in the world, with 18,700 gas and service stations worldwide, hired the public relations professionals Ogilvy & Mather to promote the slant that climate change is not the fault of an oil giant, but that of individuals.*
>
> *"It's here that British Petroleum, or BP, first promoted and soon successfully popularized the term 'carbon footprint' in the early aughts. The company unveiled its 'carbon footprint calculator' in 2004 so one could assess how their normal daily life—going to work, buying food, and (gasp) travelling—is largely responsible for heating the globe."*[3]

It can be exhausting to shoulder the weight of all the things that are not right in the world today, and to believe that it's up to us as individuals to make the necessary changes when larger voices have significantly more power and can act with impunity. Yes, our individual choices can effect changes, but they are much more impactful if we work collectively to change the system.

If I decide that I will only purchase wool/yarn that has a fully traceable value chain, I will make a small impact and will feel confident that my dollars are being spent on materials and businesses that I trust.

But if we collectively start demanding that companies share the traceability information about their yarn/wool/textile products on their labels, then this sort of transparency will become common practice and will improve the industry as a whole. As consumers we need to hold the companies that we purchase from accountable and demand clear and transparent labelling on all our products. This is being done already by some clothing and yarn companies using blockchain technology which tracks and guarantees traceability along the entire value chain.[4]

3 Mark Kaufman, "The Carbon Footprint Sham," *Mashable*, https://mashable.com/feature/carbon-footprint-pr-campaign-sham.

4 I recommend looking at Nativa as a practical example of how to use this technology: https://www.nativapreciousfiber.com/blockchain/Reformation_2021_QR.

We can also demonstrate the positive changes we want to see in the industry by supporting those companies and farms that are already practising a traceable model, by shopping with them and keeping our money in our local communities. Purchasing a sweater's quantity of yarn from a local farmer will have a big impact on their long-term viability. That money goes directly to caring for sheep and stewarding land and supporting small mills. For every dollar spent in a local business, 66 percent of it will stay in our local communities.[5] If you have the opportunity to purchase yarn from a local farmer, mill, or community-based yarn business, your money will circulate through the community and have an even larger impact.

I encourage you to reach out to your local fibre farmer and visit their farm or mill. Make the effort to get to know them and how they farm and care for their animals and land, and maybe even offer to skirt some fleeces with them.

Advocate for Change

There are many paths to changing our wool industry, and there are many different players who need to take action. In the last decade or so, there have been renewed conversations and calls to action to revitalize the industry, and many new (and old) actors who have added their voices to the chorus. The Campaign for Wool Canada, the multiple Fibreshed organizations across the country, The Canadian Wool Collective, and numerous farmers, mill owners, fashion designers, wool advocates, entrepreneurs, and authors have started to speak up. I am hopeful that we are contributing to a large, unified voice that clearly articulates that wool is important and valuable and that we want a better system for all involved. Although we might not always agree on how to get there, I hope that we all will continue pushing for change.

In my relatively short time as a sheep farmer, mill owner, and wool advocate I have tried to understand the history and present state of our industry so that I can push for a future that is better. I would like

5 "The Power of Purchasing: The Economic Impacts of Local Procurement," Canadian Federation of Independent Business, 2023.

> *The first step in growing the Canadian wool industry is to revalue wool.*

to put forward my vision for our industry—one that prioritizes community economic development and employment, ecological regeneration, regionally specific adaptation and resources, and a fully traceable value chain.

The first step in growing the Canadian wool industry is to revalue wool. The greatest barrier to this is mindset and compensation. For too long, farmers have not been compensated for their wool in a way that incentivizes them to reduce contamination. Managing their wool has become "cost of doing business" for many producers who have no motivation to change. The industry needs to come together to increase pricing per pound for Canadian wool and commit to providing situation-appropriate training on wool handling, to pair with the assurance of higher per-pound rates.

I believe that wool pellet fertilizer is the first step in revaluing wool. Wool pellets are a proven effective soil amendment that offer slow-release nitrogen and other nutrients and decrease the need for irrigation by increasing water-holding capacity within the soil;[6] they also improve aeration and can limit pests like slugs. Pelletizing wool requires less infrastructure than all other forms of wool production because it does not require scouring (washing).

If we worked together to establish wool pelleting operations in every region across Canada (many already exist), we can limit the decentralized movement and associated GHG emissions of our current wool transport system. It is feasible to pay, at minimum, one dollar per pound for pelleting wool, which far exceeds the current rates most farmers are paid for their wool. The challenge is to market this much

6 T. Bradshaw and K. Hagen, "Wool Pellets Are a Viable Alternative to Commercial Fertilizer for Organic Vegetable Production," *Agronomy*, 12(5):1210 (2022), https://doi.org/10.3390/agronomy12051210.

wool pellet fertilizer. I propose that we collaborate with city, provincial, or regional planning departments. These departments manage all the green spaces and gardens in our cities, and every year many apply synthetic fertilizers and other products to assist the growth and water retention of our city green spaces. Imagine if all city green spaces, planters, and garden beds were fertilized with wool pellets that had been grown within a short distance of those cities and towns. The impact of this could be immediate:

- Natural fertilizer would replace synthetic fertilizers
- Nutrients would be cycled from local soils back into those soils in a circular system
- Local sheep farmers would be compensated adequately for their wool
- Jobs would be created in the region by enhanced pelleting operations

Governments are notorious for moving slowly, but generally our municipal governments are the most accessible levels of government for collective citizen action. If all the knitters and crocheters and weavers across Canada began speaking to elected municipal officials about the benefits of wool in our soils and the potential ripple effects for regional sheep farmers and textile systems, we could make a change within the current term of government.

But my vision doesn't stop with wool pellets. As commercial producers begin to see the financial benefit of wool, we should work with provincial and national sheep organizations to offer wool-handling training that can promote the cleanliness of wool with the promise of higher rates when the wool is sold. We need to stop asking farmers to work harder to keep their wool clean, without giving them the assurance of adequate compensation for that work.

The next step is to map out and identify regionally specific resources and gaps in the wool value chain. The sheep breeds, management styles, and physical resources look very different in Atlantic Canada than they do on the Prairies. We need to identify and capitalize on these regional differences to build up and produce appropriate infrastructure and products where they are needed.

For example, many of the sheep farms in the Prairies are large ranches or pasture-based grazing operations whose sheep are generally the fine-wool breeds that are appropriate for textiles, like Rambouillet, Targhee, Corriedale, et cetera. These breeds are not often seen in wetter climates like those in our coastal regions. It would make sense to develop a smaller, fine-wool-specific infrastructure hub for this region, different from what would be needed on the coasts.

Currently, there is one large scouring facility in Canada, located in Ontario. As I discussed earlier, scouring is a major bottleneck in the domestic processing of the Canadian wool clip. Manitoba has a lot of water resources, so we could invest in building a regional scouring facility that has the capacity to scour all the wool grown in Alberta, Manitoba, and Saskatchewan each year. Eliminating this key bottleneck in wool processing could remove the barriers for new infrastructure to develop. A fine-wool spinning mill could be run on a cooperative basis to swiftly and efficiently process all the fine wool in the region and could work in partnership with a weaving or knitting mill and local designers to produce 100 percent Canadian wool toques or sweaters.

When we approach this kind of challenge in a smaller, more regional manner, we can play with actual numbers. I live in Manitoba and we grow approximately 51,000 pounds of wool a year. If it was fine wool and we had adequate infrastructure, we could turn that 51,000 pounds into 14,285 wool sweaters or 115,000 wool toques or 100,000 pairs of wool socks.

Or perhaps we might assess that the wool in Manitoba is better suited to household textiles. From the 51,000 pounds of wool, we could produce 16,000 wool pillows or 6,250 wool duvets or 2,700 carpets. Obviously, this won't meet the needs of our entire population—at least not at first—but this thought experiment can provide inspiration for what could actually transpire if we developed regional processing facilities.

The greatest barriers to plans like these are lack of leadership and capital. Infrastructure is costly and doesn't already exist in Canada, but we must start somewhere. Perhaps wool pellet operations could generate enough revenue to invest in more processing equipment. Perhaps CCWG stakeholders could push the co-op to invest their profits into regional processing infrastructure. Maybe you are reading this and you work in government or industry and you could advocate for wool processing to be a priority for grants and subsidies. Perhaps you are someone with the means to invest in a start-up regional scouring facility or wool insulation operation. Or perhaps you are someone who could buy a skein of yarn from a local fibre farmer, or even just send them an email thanking them for their commitment to Canadian wool.

Whatever role each of us plays in the Canadian wool industry, we can all make a change, one stitch at a time.

APPENDIX

These are the expanded fixed and variable costs for the three sheep-farm scenarios introduced on page 34. They are based on real numbers, and on interviews with farmers.

Scenario 1

Cost of Production for 1 Sheep for 1 Year

Fixed Costs:

- Maintenance of buildings/fences and utilities: $12.00
- Property costs (insurance, mortgage, and property tax): $42.65

Variable Costs:

- Feed (hay and other rations) from October–June: $72.55
- Grain, minerals, and salt: $24.30
- Fecal samples, veterinary, and healthcare: $6.50
- Bedding: $5.50

Total Fixed Costs: $54.65
Total Variable Costs: $115.65
TOTAL: $170.30

Amount of wool from 1 sheep in 1 year: 5 lb. coloured wool

Potential Revenue from 1 Sheep:

- 5 lb. of wool sold raw to local mill: $40.00 or
- 5 lb. of wool sold raw to handspinner: $85.00 or
- 5 lb. of wool sold to CCWG: $6.25

Scenario 2

Cost of Production for 1 Sheep for 1 Year

Fixed Costs:

- Maintenance of buildings/fences and utilities: $8.00
- Machinery Maintenance: $10.00
- Property costs (insurance, property tax and insurance): $65.00

Variable Costs:

- Feed from October–June: $83.69
- Grains, minerals, and salt: $56.14
- Fecal samples, veterinary, and healthcare: $7.00
- Bedding: $4.50
- Registrations and transfers of purebreds: $9.66

Total Fixed Costs: $83.00
Total Variable Costs: $167.99
TOTAL: $250.99

Amount of wool from 1 sheep in 1 year: 8.5 lb. fine wool

Potential Revenue from 1 Sheep:

- 8.5 lb. of wool sold raw to local mill: $26.78 or
- 8.5 lb. of wool sold raw to handspinner: $68.00 or
- 8.5 lb. of wool sold to CCWG: $10.47

Price for commercial market lamb (meat): $280.00
Price of breeding stock (if a ewe had 1 lamb and it was sold as breeding stock): $11.00

Scenario 3

Cost of Production for 1 Sheep for 1 Year

Fixed Costs:

- Maintenance of buildings/fences and utilities: $11.20
- Property costs (insurance, property tax, and insurance): $37.11

Variable Costs:

- Feed (homegrown and purchased): $70.12
- Grains, minerals, and salt: $30.99
- Fecal samples, veterinary, and healthcare: $4.59
- Bedding: $11.52

Total Fixed Costs: $48.31
Total Variable Costs: $117.22
TOTAL: $165.53

Total amount of wool from 1 sheep in 1 year: 8 lb. medium wool

Potential revenue from 1 sheep:

- 8 lb. of wool sold raw to local mill: $25.00 or
- 8 lb. of wool sold raw to handspinner: $32.00 or
- 8 lb. of wool sold to CCWG: $3.60

Price for ewe sold to commercial meat market: $180.00
Price of breeding stock (if a ewe had 1 lamb and it was sold as breeding stock): $200.00

RESOURCES

BIBLIOGRAPHY

Author's Note

Bacon, Ashley. "Ontario wool producers focused on the positive amid U.S. tariff threat." CTV News. March 6, 2025. https://www.ctvnews.ca/kitchener/article/ontario-wool-producers-focused-on-the-positive-amid-us-tariff-threat/.

Introduction

Bjersgo, Eric. "Market Report." *Canadian Cooperative Wool Growers*. Fall 2020. https://wool.ca/page/wool-market-reports.

Lu, Sheng. *FASH455 Global Apparel & Textile Trade and Sourcing*. © 2012–2025. Department of Fashion & Apparel Studies, University of Delaware. https://shenglufashion.com/.

"Responsible Business Conduct in China Textile Industry." *International Labour Organization*. ILO. January 23, 2023. https://www.ilo.org/resource/project-document/responsible-business-conduct-china-textile-industry.

Holt, R.R.D. "Introduction to Superwash Wool." *Journal of the Society of Dyers and Colourists* 91 (1975): 38-44. https://doi.org/10.1111/j.1478-4408.1975.tb03228.x.

Chapter 1: Growing Wool

Government of Canada. *Number of sheep and lambs on farms (× 1,000)* Table 32-10-0129-01. Statistics Canada. Release date: February 25, 2025. https://www150.statcan.gc.ca/t1/tbl1/en/tv.action?pid=3210012901.

Government of Canada. *Market Overview—Canadian lamb/sheep meat and product trends*. Agriculture and Agri-Food Canada. https://agriculture.canada.ca/en/international-trade/market-intelligence/reports/customized-report-services-market-overview-canadian-lambsheep-meat-and-product-trends#b.

Government of Canada. *The Wool Textile Industries 1953*. Dominion Bureau of Statistics Industry and Merchandising Division Vol.2 - Part VII—D-I. Government of Canada Publications.

Government of Canada. *Sheep statistics, quantity of raw wool purchased, used on farms, price and farm value*. Table 32-10-0142-01. Statistics Canada. Release date April 19, 2024. https://www150.statcan.gc.ca/t1/tbl1/en/tv.action?pid=3210014201.

Government of Canada. *Biannual Livestock Survey*. https://www23.statcan.gc.ca/imdb/p2SV.plFunction=getSurvey&SDDS=3460. Statistics Canada. Accessed October 2024.

Koronios, Eva. "Fabric, Craft & Sewing Supplies Stores in Canada Doing Sew Well: Increased operating efficiency among leading operators is anticipated to preserve industry profit." *Craft Industry Alliance*. May 2020. https://craftindustryalliance.org/wp-content/uploads/2021/11/May-2020-Fabric-Craft-Sewing-Supplies-Stores-in-Canada-Industry-Report.pdf.

Government of Canada. Canadian Importers Database..https://ised-isde.canada.ca/app/ixb/cid-bdic/productReportHS10.html?hsCode=5109900000&wbdisable=true Accessed October 2024.

Government of Canada. Farm Credit Canada. *Historic FCC Farmland Values*. Government of Canada Publications. March 13, 2023. https://publications.gc.ca/collections/collection_2024/fac-fcc/CC213-7-2023-eng.pdf.

Government of Canada. Farm Credit Canada. *2023 FFC Farmland Values Report*. Government of Canada Publications. March 12, 2024. https://www.fcc-fac.ca/en/knowledge/economics/farmland-values-report.html#:~:text=Farmland%20continued%20steady%20climb,2021%20and%205.4%25%20in%202020.

Qualman, Darrin, Annette Desmarais, Andre Magnin, and Mengistu Wendimu. "Concentration Matters: Farmland Inequality on the Prairies." *Canadian Centre for Policy Alternatives*. 2020. https://policyalternatives.ca/sites/default/files/uploads/publications/Saskatchewan%20Office%2C%20Manitoba%20Office/2020/11/Farmland%20Concentration.pdf.

Simpson, W.S. and G.H. Crawshaw, *Wool: Science and Technology*. Woodhead Publishing Ltd. 2002.

Hergenham, R. "Nutritional Determinants of Wool Quality and Quantity." *Woolwise. The Australian Wool Education Trust*. 2014 https://www.woolwise.com/wp-content/uploads/2017/07/WOOL-300-300-14-T-04.pdf.

"Growing the Next Crop of Canadian Farmers." *Action Canada*. https://actioncanada.ca/wp-content/uploads/2021/06/TF-3-WEB-ENG.pdf.

Government of New Zealand. Manatū Taonga—Ministry for Culture and Heritage. Godfrey Bowen. https://nzhistory.govt.nz/people/godfrey-bowen Updated 8-Nov-2017.

"Biological Harvesting Update." *Australian Wool Innovation Ltd*. https://www.wool.com/news-events/news/biological-harvesting/ Accessed October 2024.

Wilkes, Stephany. *Raw Material: Working Wool in the West*. Oregon State University Press, 2018.

Chapter 2: Shipping Wool

Government of Canada. *Greenhouse Gas Emissions*. https://www.canada.ca/en/environment-climate-change/services/environmental-indicators/greenhouse-gas-emissions.html #transport. Accessed September 2024.

Plumptre, Bora and Eli Angen, and Dianne Zimmerman. "The State of Freight." *Pembina Institute*. June 2017. www.pembina.org/pub/state-freight.

"Annual Report 2018." *Canadian Cooperative Wool Growers*. February 2018. https://www.wool.ca/images/uploads/files/2018-Annual-Report.pdf.

Chapter 3: Processing Wool

Burgess, Rebecca. *Fibershed: Growing a Movement of Farmers, Fashion Activists, and Makers for a New Textile Economy*. Chelsea Green Publishing. 2019.

Government of Canada. Minister of Supply and Services Canada. McCullough, A.B. *The Primary Textile Industry in Canada History and Heritage*. Government of Canada Publications. 1992. https://publications.gc.ca/site/eng/9.609575/publication.html.

Government of Canada. Department of Agriculture. *Census of Canada 1870-71*. Vol. 3, 1873. Government of Canada Publications. https://publications.gc.ca/collections/collection_2016/statcan/CS98-1871-3.pdf.

Government of Canada. *Census Report of The Canadas 1851-2. Vol. 7, Lower Canada—Return of Mills, Manufactories.* Government of Canada Publications. https://publications.gc.ca/collections/collection_2012/statcan/rh-hc/CS98-1851-2-eng.pdf.

Government of Canada. *Report of the Royal Commission on the Textile Industry.* Turgeon, Justice W. F. A. 1938. *Government of Canada Publications.* https://epe.lac-bac.gc.ca/100/200/301/pco-bcp/commissions-ef/turgeon1938b-eng/turgeon1938b-part1-eng.pdf.

Porritt, Edward. "Canada's National Policy." *Political Science Quarterly.* June, 1917, Vol. 32, No. 2 (June 1917), pp. 177-208. https://www.jstor.org/stable/2141729.

"A Chronology of the Textile Manufacturing Industries of Canada, 1883 – 1983." *Canadian Textile Journal.* June 1983.

Wyman, Diana. "Trade Liberalization and the Canadian Clothing Market." *Canadian Economic Observer.* December 2006.

Bloskie, C. "Canada's Textile and Clothing Industries." *Canadian Economic Observer.* March 2005.

Scott, Shirley A. *Canada Knits: Craft and Comfort in a Northern Land.* McGraw-Hill Ryerson Ltd. 1990.

Briggs & Little. https://briggsandlittle.com/history/ Accessed October 2024.

Olsen, Sylvia. *Unravelling Canada: A Knitting Odyssey.* Douglas & McIntyre. 2021.

Manitoba Co-operator Staff. "Long guns and long johns." Oct. 22, 1953. Our History. https://www.manitobacooperator.ca/country-crossroads/our-history/our-history-october-1953/.

Textile Exchange. *Responsible Wool Standard 2.2.* Copyright 2021. https://textileexchange.org/app/uploads/2020/08/RAF-101a-V2.2-Responsible-Wool-Standard.pdf.

Chapter 4: Processing Wool Internationally

Government of Canada. Department of Trade and Commerce. *Textile Industries in Canada, 1941*. Dominion Bureau of Statistics. Government of Canada Publications. 1943. https://publications.gc.ca/collections/collection_2018/statcan/34-209/CS34-209-1943-eng.pdf.

Chapter 5: Marketing Wool

Government of Canada. *Direct payments to agriculture* producers ×*1,000*. Table 32-10-0106-01. Statistics Canada. February 28, 2025. https://www150.statcan.gc.ca/t1/tbl1/en/tv.action?pid=3210010601.

Conclusion

"A new textiles economy: Redesigning fashion's future." *Ellen MacArthur Foundation*. November 27, 2017. https://www.ellenmacarthurfoundation.org/a-new-textiles-economy.

Textile Exchange. *Materials Market Report*. 2024.

Kaufman, Mark. "The Carbon Footprint Sham." *Mashable*. https://mashable.com/feature/carbon-footprint-pr-campaign-sham.

"The Power of Purchasing: The Economic Impacts of Local Procurement." Canadian Federation of Independent Business. 2023.

RECOMMENDED READING

Books

Shuman, Michael H. *The Local Economy Solution*. Chelsea Green Publishing. 2015.

Klepp, Ingun Grimstad, editor. *Local, Slow and Sustainable Fashion: Wool as a Fabric for Change*. Routledge, 2022.

Segal, Alvin Cramer. *My Peerless Story: It Starts with the Collar*. McGill-Queen's University Press. 2017.

Biggar, E.B. *Canada's Wool and Woolens: The Problem of Clothing the Canadian People with Canadian Wool Manufactured by Canadian Woolen Mills*. Forgotten Books. 2012.

Rutt, Richard. *A History of Hand Knitting.* Interweave Press Inc. 1987.

Parkes, Clara. *Vanishing Fleece: Adventures in American Wool.* Abrams Press. 2019.

Magazines

Slone, Isabel. "Why isn't more clothing made in Canada?" *Maclean's.* July 21, 2021. https://macleans.ca/society/why-isnt-more-clothing-made-in-canada/.

White, Tamara and White, Char. "The Cost of Farm Yarn." *Taproot.* Issue. 36. November 2019, pp. 12-19.

Canadian Government Resource

Government of Canada. Department of Trade and Commerce. *Textile Industries in Canada, 1951.* Dominion Bureau of Statistics. Government of Canada Publications. 1953. https://publications.gc.ca/collections/collection_2018/statcan/34-209/CS34-209-1953-eng.pdf.

ACKNOWLEDGMENTS

After my first book, *Sheep, Shepherd & Land*, was published I recognized all the people that I did not acknowledge in writing. I vowed that for my next book I would be diligent in keeping a list of everyone who had a hand in making it happen. Sadly, I did not keep that list and I'm sure these acknowledgments will be incomplete. This book has challenged me and pushed me in ways that were surprising. Many have lent a supportive ear or a thoughtful response as I navigated the complex world of globalized production and trade and wrestled with the challenges facing our textile industry. I am deeply grateful for those who have been a part of my journey, both in farming and wool advocacy, and any omissions here are inadvertent.

Thank you to all my supporters for pre-ordering this book and bringing it to life. Thank you for encouraging me and showing up for me, whether it was a comment on social media or an email to just say that you care—it means the world.

This book could not have happened without the competent and passionate team of people that have worked on it and who sprung into action to release it months earlier than originally planned. To Lisa Friesen, who did layout and design and probably knows more about sheep and wool than she ever intended—thank you for always capturing my vision with visuals and graphics. Michelle Woodvine, thank you for your copyediting and fact checking and for seeing the mistakes that I could not. Thank you, Kim Werker, for enthusiastically saying yes to publishing this book, for thoughtfully and critically examining the manuscript and challenging me to make it better, and for being so passionate about changing the world and inspiring those around you to do the same.

Thank you, Clara Parkes, for writing the foreword and for inspiring me to dig deeper into wool.

Thank you to all the sheep farmers, shearers and industry folks who have generously shared knowledge and answered my questions and shepherded me through my own journey of wool—Graham Rannie, Gerry Oliver, Lisa Surber, Margaret Brook, Russel Eddie, Chloe Bermejo, Stacey Rosvold, Ursina Studhalter, Romy Schill, Sherry and Harold Bosma, Ingrid Stalman, Randy Eros, Barb Mulock, Linda Glowacki. Thank you to the mill owners and those involved in the wool industry who not only answered my calls but were willing to share openly and honestly about their experiences, history, business models, challenges, and their true cost of wool production—Leah Little, Maddy Purves-Smith, Wave Weir, Brittany Vogt, Kim Doherty-Smith, Catherine Knuttson, Emily McIvor, Karri Munn-Venn, and Shirley Harvey.

Thank you to Jane Underhill, who consulted on this book. Thank you for your fact checking and perspective on the Canadian wool industry (past, present and future). I have learned so much from the work you do.

Thank you to Molly McCullough, who was willing to go through boxes in her basement and send me a signed copy of her dad's book, *The Primary Textile Industry in Canada History and Heritage* (National Historic Sites Publications, Environment Canada, 1992). That serendipitous book started me down the rabbit hole of our Canadian wool history.

Thanks to those who read early versions of this book, or parts of it, through Patreon and provided feedback—Tara Klager, Victoria Radauskus, Christel Lanthier, Christina Lees, Ruth Stubens, and Brittany Casavant. And to those who provided blurbs for the book (under a tight turnaround time)—Alexa Ludeman, Leah Little, and Caroline Sommerfeld, thank you!

Thank you to my friends who held space for me to work through the challenging parts of this book—you held me up while I was discouraged, and let me talk endlessly through my many doubts. Kat Funk, Bethany Booy, Pamela Delisle, Crystal Salkeld, and Nicole Vechina, you somehow knew when to listen, when to give feedback and when to bring out the chocolate.

Christel, thank you for the endless conversations about the wool industry, sheep raising, colonization and reconciliation, textile industry, business planning, and general badass shepherdass things. You are my sounding board and my inspiration in this business and I'm so grateful for you.

Luke, Bo, and Josh, my loves and my heart, this book would not be complete without your ongoing support. Luke, I know you didn't build the cabin for me to write in, but you didn't complain when I took it over. I know my run-on sentences are maddening, but you never fail to catch them and insert the punctuation. I know my half-baked, far-fetched plans can be exhausting, but you have never tried to corral my spirit—thank you for being the one that brings my visions to life.

ANNA HUNTER is a first-generation sheep farmer and wool mill owner in Eastern Manitoba, Treaty One Territory. She is passionate about building community and connecting rural fibre farmers with urban consumers, fibre artists, and crafters. Anna believes that regenerative agriculture and climate-beneficial food and clothing are integral to moving forward as farmers, fibre artists, and Manitobans, and she is passionate about growing a vibrant Canadian wool industry. Her first book, *Sheep, Shepherd and Land*, came out in 2023. Learn more about Anna and her farm and wool mill at longwayhomestead.com.

PUBLICATIONS

Nine Ten publishes non-fiction books by Canadians on topics related to art, craft and creativity. We get very excited about books that explore a topic or topics deeply, written by authors who know a whole lot about something most folks know very little about. It's not that our goal is to shine light on the obscure, but rather to serve readers who hunger for books that scratch beneath the surface of the topics they're passionate, nerdy, geeky and obsessive about. We print our books in Canada, on paper that has as much recycled content as we can afford (usually, that's 100%).

Find more fascinating books at shop.ninetenpublications.ca, at your local library, or at your favourite craft shop or bookstore.

ALSO FROM AUTHOR ANNA HUNTER

Sheep, Shepherd & Land:
Stories from Small Farms Reinvigorating Canadian Wool

"... a beautiful book celebrating the diversity of Canadian wool and the tenacity of Canadian sheep farmers."

—knitty.com

Across Canada, small-scale sheep farmers are on a mission to reinvigorate the domestic wool industry, raising sheep breeds that thrive in their regional climate and land base. Learn about farmers' motivations, their dedication to climate-forward practices, and their passion for developing the infrastructure to ensure their wool is processed, milled, and used by businesses and hobbyists locally. Journey alongside sheep farmer and wool educator Anna Hunter to explore the unique characteristics of the sheep breeds being raised from British Columbia to the Maritimes, and how the qualities of the wool from each is best suited to make items from garments to cold-weather accessories to home goods.